Rediscovering Vinland

*To
Barbara Fontaine
Long Time Friend*

Rediscovering Vinland

Evidence of Ancient Viking Presence in America

Fred N. Brown, III

iUniverse, Inc.
New York Bloomington

Rediscovering Vinland
Evidence of Ancient Viking Presence in America

Copyright © 2007, 2009 by Frederick N. Brown, III

All rights reserved. No part of this book may be used or reproduced by any means, graphic, electronic, or mechanical, including photocopying, recording, taping or by any information storage retrieval system without the written permission of the publisher except in the case of brief quotations embodied in critical articles and reviews.

iUniverse books may be ordered through booksellers or by contacting:

iUniverse
1663 Liberty Drive
Bloomington, IN 47403
www.iuniverse.com
1-800-Authors (1-800-288-4677)

Because of the dynamic nature of the Internet, any Web addresses or links contained in this book may have changed since publication and may no longer be valid.

The views expressed in this work are solely those of the author and do not necessarily reflect the views of the publisher, and the publisher hereby disclaims any responsibility for them.

ISBN: 978-0-595-43680-4 (pbk)
ISBN: 978-0-595-88008-9 (ebk)

Printed in the United States of America

Contents

Foreword xi
Preface xvii
Acknowledgments xix
Introduction xxi
Timeline of Significant Events xxvii

PART I: VIKINGS 1

We look deeply at a people viewed universally as primitive barbarians. We find, instead, a peaceable and stable culture of farmers, fishermen, and seamen. Advanced for their day, they bequeathed to the Western world a refined sense of law and formed the first and longest lasting of democratic governments.

Where Do We Start? 3
A Brief History of Vikings 8
Viking Navigation 13
L'Anse aux Meadows: A Proven Norse Settlement in America 17
The End of the Greenland Vikings 22
What the Vikings Left Behind 27

PART II: THE VINLAND SAGAS 37

The Vinland Sagas entire and as far as we know, exclusive. "You can't enjoy the game without a program." This narrative is the basis of our program.

Introduction to the Sagas 39

The Saga of Bjarne Herjolfson 42
The Saga of Leifur Eiricksson 44
The Voyage of Thorvald Eiricksson 51
The Saga of Thorstein Eiricksson 55
The Settlement Expedition 58
Vinland Landing 61
Thorhall's Departure 64
The Skirmishes at the Settlement 68
The Saga of Freydis Ericksdottir 73
Conclusion 77

PART III: THE VINLAND SAGAS AS A GUIDE 81

An elusive historical "target," aimed for by many for centuries. A "voyage" that commenced as a lark but developed into a serious endeavor.

The Voyage of Wave Cleaver 83
The Coast of Vinland Explored from Space and Approached Anew 92
The Coast of Vinland: Possible Locations of Leifsbudir, Hop,
    Straumney and Crossannes 107
The Coastline as Developed by the Voyage of Wave Cleaver 118

PART IV: OBSERVATIONS OF NARRAGANSETT AMERINDS BY EARLY VISITORS TO AMERICA 123

We consider the idea that perhaps Vikings influenced native aborigines during their supposedly "brief" settlement, a miracle of locating two perfect resources—intelligent, and educated, precise men with no ax to grind.

Giovanni da Verrazzano, 1524 125
Narragansett Tribal Lands When Roger Williams Arrived in
    1635 132

PART V: NARRAGANSETT DIFFERENCES 143

We detail previously unexamined recordings of these people and find them unique among all other American Indian tribes. We introduce recent archaeological evidence that this uniqueness has its roots in an archaic incursion with dating coincidental with that of the Vinland voyages.

Physical Differences 145

Language Differences 150

Number Differences 161

Burial Differences 166

Sexual and Sanction Similarities 171

Water and Land Use Differences 177

Conclusions about Narragansett Differences 186

PART VI: NARRAGANSETT GENETICS SHOW EUROPEAN ANCESTRY 191

A detailed discussion of the genetic factor; how and why this shows that Narragansetts and Vikings were at one time intimate.

Narragansetts Have Pre-Columbian Genetic Differences 193

Narragansetts Resisted Tuberculosis Because of Previous Exposure 195

Tuberculosis in America 199

Would DNA Tests Determine Viking Descent? 213

Who Else Could Have Interbred with Narragansett Indians? 217

PART VII: ADDITIONAL CLUES TO NARRAGANSETT DESCENT FROM VIKINGS 227

Suspicion of a single, seemingly isolated, factor leads us to consider possible alternative indications of disease immunity.

Investigation and Commentary for Additional Clues to the
   Immunology Factor 229
Initial Colonial Settlement...232
Providence Settlement 235
Intimacy of the Races..239
We Hear Again from John Winthrop, Jr...242
Narragansett Indians Decline..245
Some Notes from Professor McNeill..248

PART VIII: "REFUGIO?" 253

In which we delve into local colonial history for additional information. We find much of incidental interest, some of which is in support of our theme. We compare the Viking landings to those of certain original "Yankee" arrivals. We find them closely comparable and we end with commentary concerning two who are certainly the most perfect resources available to scholars of this subject.

Was Jireh Bull's Property a Viking Site?...255
"Refugio" and "Norman Villa" 268
Roger Williams and a Viking Legacy for America 271

Afterword 279
Sources 281
Index 289

Cover photograph, courtesy Erick O. Brown

This is a most informative photograph, which compares several points of information from the Vinland sagas regarding Leifsbudir and Hop. It was taken along the sandbar at Narrow River, Rhode Island, in an ENE direction on January 10, 2007. That day is only some 19 days from the winter solstice (shortest day of the year).

According to the sagas, in Vinland at "Dagmalastad" (8:00 a.m.) the sun was "up," as it is here at 7:00 a.m., nearly an hour earlier. This is one of the more potent indices that hint of a far southerly destination for the Vinland explorers. (This nineteen day and one hour difference probably duplicates the conditions of AD1000 on December 22/23 in 2007.)

The picture also shows conditions there at low tide and something of the geomorphology of the river mouth and sandbar, which, according to the sagas, lay "… directly across the mouth of the river." Not very visible are the strong outward currents and the fact that on the left (inland) side, water depth is some four feet while on the right (sea, or approach side), it is only near ankle-deep. One might detect the strong outward current just off the tip of the bar. Passage into Narrow River cannot be made except at two periods of the day when high tide/slack water conditions prevail. One of these periods would normally occur at night, so that leaves but one time of the day for entry. Exit might have a somewhat longer window of opportunity by being able to ride the current as long as there was a depth of water outside.

If a sailing ship had the good fortune to approach at one of those times, maneuvering would entail a full 180-degree turn, and might encounter an adverse wind at some times. However, rowing is feasible in calm conditions. From the sea, it is an unappealing sight and, except for the enticing view of expansive strands and inviting inland conditions, would likely be bypassed.

At low tide, it seems narrow and challenging, while at high tide it seems so broad as to invite entry. The main problem is offshore where the outlying rocks force a maneuver close to the strands. I have tried this entry four times with small craft and grounded twice even though I knew where the channel was. Once was outside and the other inside into the "false" channel that runs along the north shore but ends disappointingly abruptly.

The view is calm and one cannot conceive of how rough conditions can be when exposed to open ocean storms. Not visible beyond the rock is a rather close-in ruin of a concrete caisson which had been the base of a lighthouse. It was constructed to survive the most severe storms imaginable, being made of cast iron and some 73 feet high. In 1924, a wave swept right over the top causing some destruction of the roof. But in 1938, a famous hurricane wrought immense damage all through the region and which I well remember. This storm swept the lighthouse entirely away along with the keeper. Moreover, very little of the cast iron wreckage was ever found. Elsewhere in this book is remarked the "boom and crash of devastation." Those familiar with the sea have great respect for these episodes when the wrath of Neptune is let loose.

In that 1938 storm, Narrow River inside this bar sustained only some storm "surge" damage and Pettaquamscutt Lake, less than a mile inside, almost none at all. The town of Narragansett Pier immediately to the south was inundated and destroyed. Providence, 30 miles north, was also inundated but the well-built structures survived even if their lower floor contents did not.

Descriptions of Leif Erickson's grounding, which might have been somewhere in this view, indicate that the "afterboat" (tender) must have been used to investigate these difficulties. Its small crew most probably would have been much intrigued ("—*they tumbled over the side because they could not wait to explore*") by the beautiful and placid river, which flowed from the inland lake just visible from this spot.

Foreword

After more than 30 years of study on the subject of the Norse Vikings, I have come to believe that they landed, among other places, in Rhode Island. This book is the story of how I came to that belief. You should, I suppose, know a little about me.

I could read well before I ever saw the inside of a kindergarten classroom, thanks to my mother and my own efforts to teach myself. My father's sometime habit was to accompany me to a library and patiently await eventual lapse of my curiosity, for I grew up in New England where frequent changes in weather enforced retirement to warm sheltered places—and libraries are ideal places for such shelter.

By age 13, I was an accomplished reader. I used libraries for refuge from weather, irate neighbors, teachers who had noticed my absence, and perhaps (dare I say it?) sometimes the police. By that time, I was also very much the "out and about" type of boy with the lively, occasionally destructive imagination typical of boys that age. I was, in fact, rapidly headed for trouble, but when that trouble arrived, it came from an entirely unexpected direction.

World War II was just getting underway for Americans and my area was alive with uniforms of servicemen who carried their risks, spirits, and diseases from distant climes with them.

A microbe escaped from a nearby naval base where an epidemic of it raged and nearly carried me out of this world. I wakened from a long coma to find myself in a world of silence. My hearing and a few other faculties, like balance, were shot to hell in a hand basket.

Until that time, I had never taken the world very seriously, but soon felt it was past time to take things seriously indeed. The world had been my oyster, so I felt, but it then occurred to me that when judging just who was an oyster, the world had more experience than I.

By necessity, I had to attend my last year of junior high school at a state school for the deaf, which was at that time a so-called "oral" school using lip-reading instead of the more common sign language skills. They had a surprisingly good library but it took a little while for me to discover it.

I had to endure a bitter experience from my position of "Johnny come lately" to attempt assimilation—unsuccessfully—into a rather closed society of people who were born deaf or lost hearing at an earlier age.

The library called and I answered, moving to that refuge and a book that saved my sanity.

The book was Mark Twain's *Roughing It* and it spoke to me of "out and about" and new worlds undiscovered. Since that time, I have been an avid reader and now find a good book easily as vivid as any motion picture or television program. However, I will not accept the view that I am a "bookworm," for I am still very much the out and about type of guy.

Perhaps I have had more experience than one might suppose for a deaf fellow. I took up target rifle shooting and progressed to participation in the highest of competitive matches, but I must tell you that at that time I wore a hearing aid, which enabled me to hear firing commands.

I motorcycled avidly for many years, using it for basic transportation in all weathers and temperatures, riding in TT (Tough Terrain) competitions, traveling safely clear across the country to learn yet another lesson of life. I found that Mr. and Mrs. "average American" are great, helpful, and welcoming in the main. I even found that the usually derided motorcycle fraternity is, in fact, made up of all sorts, friendly, accommodating, and lawful people—a great bunch to be out and about with. I managed somehow to raise, rebuild, and live aboard (year-round for nearly a decade) a 38-foot powerboat of a type similar to a lobster boat. Another lesson learned, for the people of the nautical community also have their nobility, fraternity, and capabilities.

I have learned to fly ultra-light aircraft but have not pursued it because of my uncertain sense of balance, which seems to promise, eventually, a very unhappy landing. I choose not to chance that promise.

"Out and about" or "on the road" for years sounds like a difficult way of life. Yet I never found it so. Meeting the world and the denizens of it—human and otherwise—gave me the knowledge that with reasonable judgment, the world is both beautiful and kind. It is often kinder than human society. The animal world is made up mostly of individuals trying mightily to avoid trouble. In that light, they are good citizens all, easily determining the distance and space required for good relations.

About the time I became again "out and about," somewhere near age 19, I had the good fortune to be able to acquire a trade as machine toolmaker in a prestigious manufacturing plant and have pursued this at many firms to the present time. Not very much the out and about sort of profession; it is, however, very challenging, interesting, and intellectually rewarding. It gives perfect balance in my own life and interests.

Along the way and by no means the least of my interests and affections and unfathomable pride, I fathered three children: Erick, Katherine, and Andy. All three have joined me at one time or another in the adventure that was to become the subject of this book. I admit them as having no faults. However, if they do, it is in them that whenever I tried to give them some fatherly advice, it transpired in less time than can be imagined that they knew more than I did!

Sometime in the mid-1970s on a cold, blustery day, I came across some information in a small city library in New England concerning Leif Erickson and the Vinland voyages that he originated. I had chanced upon this issue at varied times of my life. Then later, while reading Skelton's *The Vinland Map and the Tartar Relation,* I fell asleep—or dozed off.

For some reason, I was startled wide-awake with the realization that when viewed in a certain sense, the map made perfect logic. While I am aware that the Vinland map is controversial, I have since discovered that it does, in truth, convey the essence of what was known of the New World in Europe from at least 1418 on.

By chance, that library had an excellent series of books on the subject and this commenced a three-year endeavor of reading every source I could find concerning that ancient discovery. There turned out to be far more information than I ever suspected on this important subject that most schools seem to treat in cavalier fashion. I was able to put together a tale, a saga previously fragmented that was at once more complex and informative than I could have imagined.

Three years was quite enough and I must say that this was intense to a degree. I "hit" just about every library between New Hampshire and New York City. It seemed to me imperative for more "out and about" activity and it happened that my youngest son was of an age to participate in such.

We acquired a small boat with outboard motor and commenced exploring—not with any intent to actually find something, but the exhilaration of having a purpose and being "out and about." There was no intent whatsoever for a serious discovery. However, one day we chanced upon a place, which, for some reason, struck me as fitting what I had read of the Leif Erickson landing described in vague terms and which following voyagers had embellished to a fairly clear description.

Often, it crosses my mind that in some way, destiny had crossed my path. I could not have recognized the site without those reading skills that enabled me to "see" a place described a thousand years ago. I could not have imagined it without those "out and about" experiences, so much a part of my life.

The entire project, while lengthy now at some 33 years, has been smitten with so many adventures of good fortune, and the meeting of key people at opportune times. So many times, I reached for books in libraries, which fell open to facts that I had been seeking. So many times, I found critical information in volumes seemingly unrelated to the subject. In all, the project has enjoyed amazingly good fortune, even if events proceeded at a pace I had wished was speedier. Things came along in their own good time and in their own sequence.

This explorer/writer was often stricken with feelings of isolation when my direction and train of thought seemed at odds with others. Even my hearing impairment became an aid, for I did not have to listen to the many who advised me that my task was impossible. I realize now that I was never alone. Those people I met along the way of life without whom I could not have accomplished this work are:

From way back, Allston and Mary Clarke and Isaac C. Hull, Rhode Islanders who were friends in need and friends in deed.

Donald and Hazel Wilcox, my nautical mentors, whose waterborne courtship was closely observed, interfered with, chaperoned, irritated, and survived by the writer. Our adventures at sea will never be forgotten, nor the result—a long and happy marriage and eight children, with any number of great-grandchildren.

My fellow explorers, friends, navigational beacons, and children who have illuminated my paths: Erick O. Brown, Katherine Jean Brown-Wing, and Andrew A. Brown, the latter of whom advanced to blue-water seamanship and is at present DCO Petty Officer aboard USCG Cutter Rush, based in Hawaii—the "blue-water" seaman I never became.

Officers of Incorporation of "The Voyage of Wave Cleaver, Inc.", Howard O. Barikmo and Rosemary Pedersen, whose help, contributions and enthusiasm were ever present and invaluable aids to the progress of the program.

I am grateful to the late Wilson W. Crook (Lt. Col., Ret. USA) whose interest in Anthropology and correspondence with me directed the train of evidence to the conclusion of this work.

Further, I am indebted to Dr. Diane Holloway for editing my manuscript and assisting me in the research for this book. Without her help, this book would not be.

I am most thankful for all, all of those who went before me in publication of their interest in Vinland. They are many and most now deceased, for the interest is ancient and legendary. I suspect with good reason that one of the earliest was Christopher Columbus himself, but many came after and all of them seemed to share a sense of history and adventure.

Two of those with whom I corresponded were Skulli Olafsson and Dr. Helgi Ingstad, both considerate gentlemen but now deceased. Another of the same sort who struck me as an enlightened gentleman of history is the late Dr. Albert G. Hahn of Boulder, Colorado.

There are many others whose list is too long to present here.

I live in debt to my grandfather, Dr. Fred N. Brown, a Rhode Island physician in General Practice and my father, Fred N. Brown II, an architect.

Finally, I like to spell Anthropology with a capital A with good reason. The world turns and we passengers ride along with the fates. A fate of many of us is the ongoing adventure and advancement of humankind.

Fred N. Brown, III

Preface

This book opens new doors to the story of where the Vikings went, what they did, and how they may have left Narragansett Amerinds in Rhode Island changed forever. I found this book stimulating, exciting, mind-expanding, and educational.

Fred Brown was excited by stories of Vikings as a child. His interest was piqued as he realized that he lived near the possible location of their landfall in America. He began to ask himself "Where is Vinland?"

He, like many others, believed that the Viking sagas might help him locate Vinland, much as Heinrich Schliemann believed that Homer's *Iliad* and *Odyssey* might help locate Troy.

Brown began to study the history of the Vikings. He wondered if the Vikings interbred with local natives in America as they had with so many other peoples across the world. He figured that if they interbred, there might be differences in Indians that could suggest where the Vikings settled. To determine this, he examined the observations of Indians by early explorers and colonists such as Giovanni da Verrazzano and Roger Williams.

Finding some odd differences, he then went one step further to look for genetic differences. He found that one Indian tribe appeared to resist diseases brought over by colonists better than others did. The Narragansett Indians of Rhode Island showed skeletal evidence of exposure to diseases such as tuberculosis. Such exposure, he reasoned, might have come about by a racial mixture with Viking Europeans who lived where tuberculosis was more common.

Systematically, Brown explored, sailed, researched, and read. Obviously, no one in the world can make a living by researching the question of where Vinland might have been. Nevertheless, for over 30 years, he has made this into one of the most interesting hobbies anyone can have. Like so many who have written about the Vikings, he drew his own conclusions from his own studies.

Fred Brown has followed in the footsteps of Samuel Eliot Morison, who made a career elsewhere but wrote arguably the best biography of Christopher Columbus ever written. Like Morison, he has taken his vocation (tool making), his avocation (sailing), and his boundless energy to investigate the route of the Vikings just as Morison investigated the route of Columbus.

Fred never let the fact that he has been deaf since 13 get in his way. He set out in his *Wave Cleaver* (his own small craft as well a pseudonym for his intellectual quest) to trace the route of the Vikings. What he found, where his research took him, and how he concluded that the Vikings reached Rhode Island is the subject of this book.

Morison took his vocation (admiral), avocation (history of Columbus), energy, and the boats of friends to do likewise. Morison's 1954 preface to *Christopher Columbus, Mariner* is excerpted here and Brown could say something similar:

> *The life and voyages of Christopher Columbus have been a hobby of mine for almost fifty years. After reading almost everything on the subject that was in print, I reached the conclusion that what Columbus wanted was a sailor biographer, one who knew ships and sailing and who had visited, under sail, the islands and mainland that he discovered ... My account is a straightforward narrative, giving my own conclusions on the numerous controversial points in the Admiral's career.... My point of view is still that of a sailor, relating the achievements of him whom I believe to have been one of the greatest mariners, if not the very greatest, of all time.*

Fred invited me to edit his book, which I have done. He did, however, decide to use his own wording for certain sections such as Part Two: The Vinland Sagas. I have enjoyed working with Fred and learning so much valuable information about the Vikings.

Diane Holloway, Ph.D. Editor. Author of *Who Killed New Orleans, Dallas and the Jack Ruby Trial,* and *American History in Song.*

Acknowledgments

We are grateful to the Museum of Art, Rhode Island School of Design for permission to publish the oil painting of Ninigret to iUniverse.com for publication in this book. In particular, we wish to thank Melody Ennis, Coordinator of Photographic Services.

Credit for the reproduction of 48.246:

> Museum of Art, Rhode Island School of Design
> Gift of Mr. Robert Winthrop
> Photography by Del Bogart

We are also grateful to NASA for providing two photographs (Cape Cod and Pettaquamscutt River and Narrow River Inlet). As per their guidelines, the use of NASA materials may not be used to state or imply the endorsement by NASA or by any NASA employee of a commercial product, service, or activity, or used in any other manner that might mislead.

We are most grateful to LuAnne Waddell for producing the map showing Jireh Bull's property and the Pettaquamscutt River Valley.

We also wish to thank Erick Brown for all other photographs including the cover.

Introduction

This book concerns the re-discovery of Vinland using the records of 200 adventurous Vikings a thousand years ago. Many have sought the lost location of Vinland. These pages narrate source documents, indicate varied landfalls along the way from Greenland to New England, and identify a site in Rhode Island, which may be the legendary Vinland. This discussion is based on the historic Icelandic sagas, analyses of early descriptions of possible Amerind descendants of the Vikings, and genetically inherited traits.

Icelandic sagas—records nearly a thousand years old—of explorations and settlement of the New World have been preserved intact. They lack cohesion and are set in medieval thought, yet dramatize and tantalizingly describe real places and events. Despite this, there is a common belief that Vinland and its Icelandic/Greenlandic explorers are mere myths.

This book chronicles 33 years of research and analyses of a place that has finally yielded multiple indications that Vinland has been found.

Genealogies of Icelanders and even some Greenlanders are extant from that day to this. The names of some Vinland voyagers are known and descendants of those very people live today, including a few in the United States itself!

People wonder what Vikings looked like, as if they had disappeared. I have photographs of some of these descendants in direct lineage who look precisely as other Scandinavians of our own times. Vinland is not old. As narrative, the famous Viking sagas tell of but yesterday and speak the same language as modern seamen of sail. They will live so long as men have a sense of adventure. News comes in every day for those willing to persist in the search for Vinland.

Join me on the good ship *Wave Cleaver*. Come aboard, all you who seek adventure of the intellect. Come aboard, all you who doubt the story, for even if this exploration fails to convince you of the location of Vinland, perhaps it will add to your library of information. We course regularly into what might seem as strange seas of the intellect. Yet those oceans are actually the well-charted and reliable published documents of science.

Formal scholarship sets great store on "reaching agreement," which policy has good and sufficient use in the schoolroom. However, agreement seldom occurs in matters about Vinland. Waiting for agreement is a luxury denied a shipmaster, explorer, or investigator whose enterprise must proceed. Union Commodore and later Admiral David Farragut's famous words: "Damn the torpedoes; full speed ahead!" come to mind as a policy of our expedition. We proceed, however, with caution, watchful for snags, flotsam and jetsam that might ensnare us.

Our charts are the scientific papers as reliable as can be had. Here, in intellectual adventures, we need not fear for life or loss, but only error. For the zealous explorer, that is a common occurrence. What matters a failed experiment; what matters a balked course? Even a disappointing destination yields information and perhaps new vistas. If we must utilize guesswork, it is only to eliminate false leads and direct us to new courses. The plenitude of debate will bring us into safe harbor at last.

Wave Cleaver is the name of an imaginary ship. Only to maintain equilibrium, we might keep in mind a wooden sailing ship of some 70–80 feet similar to Viking style. That successful design is still used today worldwide, and is still seen in lifeboats aboard large passenger liners. The actual research vessel used in my explorations was but 14 feet long—but this was chosen for handiness in my in-shore investigations.

You will discover here the rich texts of the Viking sagas collated from scattered sources into a comprehensive narrative. This will open new vistas as to where and how the skilled seamen crossed the seas to America at a time when few sailors left the sight of land. Those parts of this book involving the sagas have been submitted to the National Library of Iceland and the Government of that oldest of Democratic Republics, and were received favorably there.

Follow as we describe where and how the Vinland explorers journeyed and exactly where they landed. The astounding revelations that result inform us that this original landing in the New World by Europeans was much more successful than heretofore supposed. They came, they settled, they left legacies. A thousand years ago, seafaring Vikings—champion explorers and traders—settled Iceland (about the size of Kentucky), then Greenland, and finally Vinland briefly.

Famed then, famous now, the best known—Leif Erickson—straightaway became legendary among his people for his heroic qualities. The name Leifur Eiricksson is the correct name of the better-known Leif Erickson. However, Eirick is a title (meaning leader) and not a name. Erick's father (Erick the Red) was Thorvald, so Leif's true name was likely "Leif Thorvaldsson."

Long winter nights of narrations inspired some to record his adventures in sagas, which we inherit today as glorious living history. These sagas tell of new lands that had much value not available in their cold northern climate. They identified this land as somewhere in North America. This was five centuries before the equally intrepid Christopher Columbus set sail on his own adventure.

In addition to learning much about the Vikings, I wanted to learn about the Indians with whom they interacted. We now evidence that they undoubtedly left progeny. Because I eventually came to believe they landed in the Newport area of Rhode Island, I turned to the writings of Roger Williams and others for the earliest description of Indians in the area that I suspected was Vinland.

Much was learned from the new novel, *I, Roger Williams* by Mary Lee Settle published by W.W. Norton Company in 2001. This book is not for everyone but is a literary masterpiece, which results from exemplary research. I looked at the work as factual while recognizing that, perhaps, there might be errors, they probably are so inconsequential as to be irrelevant.

That led me on a merry chase through a variety of anomalies about the Narragansett Indians who were located where I had placed the Viking landfall. I am getting a bit ahead of my story but I will summarize my 30 years of research by describing a talk I recently gave.

On January 10, 2003, I addressed this subject to a local chapter of an international intellectual society (Mensa) to which I belong. I prepared the following synopsis.

> *Christopher Columbus had awareness of the geography of the North Atlantic Ocean and his entry into the Portuguese maritime service was at a time when the Kings of Portugal and Denmark were in treaty alliance to explore west of Iceland. He also recorded that in the month of February 1477, he was at a place "100 leagues" (370 miles) west of Iceland and so far north as 73 deg. N. He may, also, have had contact with Scandinavian Vikings at that locale or in transit via Iceland and he may have had knowledge of the Vinland sagas.*
>
> *The declining European Greenland colony was representative of the high cultural order of North European peoples. "Northmen" were an over-maligned and grossly misunderstood people. Iceland was the first of the democratic republics originating coincidently with the Vinland Voyages. Vikings and seamen developed from seafaring people of the Baltic environs with ships as capable as many modern sailing vessels. Modern science has shown a high proportion of Turkish genes in the modern Icelandic population—the distance from Iceland to Constantinople being a fifth the distance around the world.*

Norse sagas are not only superior literature but also are generally accurate against recorded genealogies. Many Vinland sagas were lost with Greenland colonies but partially survived in Icelandic literature and were much more detailed than is commonly known.

I have collected the previously scattered sagas into narrative form enabling an analysis of Vinland and the separate landfalls of seamen. In fact, this led to the realization that there were four separate expeditions to America. One ship traversed three times or more; one individual crossed twice; and two Norse children were born in the New World 500 years before Columbus. There were three Vinland settlements described in the sagas and another landfall so detailed as to help identify sites. There were some 18 landfalls in the New World and the sagas describe a milder climate and a solar observation indicating a locale south of 45 degrees north.

Capabilities of the ships aid in describing Cape Cod environs as Leif Erickson's Vinland. In the meantime, scholars and archaeologists at the known Viking site at L'Anse aux Meadows (Newfoundland) have concluded that the true Vinland is actually further south of that area.

The description of Leifsbudir possesses at least five criteria for the river mouth alone and at least 12 criteria for inland conditions plus one "key" criterion—a topographic feature large enough to still exist or to have its removal historically documented.

The Narrow River/Pettaquamscutt River complex approaches to Narragansett Bay in the State of Rhode Island meets all the criteria established above. More than 20 other serious studies have speculated that Leifsbudir was in New England and most others consider Karlseffni's Hop in New England or further south. Artifacts throughout New England have been discovered since colonial times, distributed in such a way as to give plausibility to this hypothesis. At least two runestones are yet accessible and observable and others are recorded.

Fortunately, the river came under scientific scrutiny because of a rare occurrence of natural topographic anomalies enabling study of certain peculiarities in Indian cultures. A comprehensive survey was made of the river complex with carbon 14 dating. Scientists expressed "astonishment" that Pettaquamscutt River alone among New England rivers demonstrated a distinct pre-historic cultural shift from former Indian waterside avoidance dwelling to a distinct and unusual waterside dwelling mode. This dynamic shift occurred "about a thousand years ago," which is coincidental with the Vinland voyages. This provided a valid approach to possible contact between two races beyond mere speculation.

This dating and this river, therefore, apparently represent the origin of the unique changes in the local tribe named Narragansetts. This tribe was recognizable at their earliest historical recording in 1524 by Giovanni da Verrazzano (see Samuel Eliot Morison's book—"The European Discovery of America").

They were described a century later (1643) by Roger Williams, an Englishman who dwelt among them and published voluminous information concerning their culture and language. He hinted strongly and in one case stated outright a perceived origin of the tribe from Iceland specifically.

> Both Verrazzano and Williams recorded the surprising nautical ability of Narragansetts, advanced abilities such as counting to high numbers in a decimal system while the English used another system, a uniform and systematic currency system, and regulated land rights and laws resulting from a hierarchal social order atypical of other Amerinds.
>
> Gregory Dexter, 1643, published Williams' book, "A Key into the Language of America," in London. Among this information is the curious observation that the terrible epidemic of 1615, which desolated and seriously depleted Indian populations all along the eastern seaboard, seems not to have affected the Narragansetts themselves.

In the 2003 address, I postulated that Pettaquamscutt/Narrow River complex was the actual landing site of Leif Erickson near 1000AD.

I also wanted to introduce the idea of genetic changes to illustrate the resistance of the Narragansetts to some illnesses. I described how antibodies develop after exposure to a disease and can leave traces in descendants showing a successful resistance to the disease in later generations. I noted that smallpox "pocks" on facial complexions (which I had observed commonly in my youth) was a successful response to that dreaded disease just as bone lesions of tuberculosis were to actual tuberculosis. While smallpox pits were an absolute success (the victim lived), the tubercular lesions were only partial, indicating that the victim was at least moderately successful in delaying the time of death; but sometimes into old age. This finding in the Narragansett population is unique and indicates an inherited genetic factor.

My synopsis was well received. I could have added some additional information such as the following.

An 1858 excavation of a 1660 Narragansett cemetery by amateur archeologist Usher Parsons, M.D., discovered an opulent burial of a Narragansett "princess." Dr. Parsons was a surgeon for the new United States Navy during the War of 1812 and he dug up an adjoining grave of her father, Sachem Ninigret. The excavation revealed several peculiarities of the population such as the opulence of the grave (indicating a high status of women in their society unusual among Amerinds elsewhere) as well as an unusual and unique precise and orderly geometric layout of the burials.

Recent excavations near the same river complex reveal a more recently excavated cemetery with graves oddly placed four abreast in a long and narrow configuration oriented southwest/northeast. Bodies were placed "almost all" on right sides, in flexed positions, and heads toward the southwest, indicating that this was a unique cultural practice unaffected by Christianity or adjacent colonists.

Of 56 bodies examined, 17 were found to have possessed tubercular exposure/resistance as revealed by lesions upon bones—and one spectacular find: a three-year-old child. Children usually die quickly upon first exposure to tuberculosis. This child's genetic resistance was strong enough to live many months—long enough to develop bone lesions! The molecular structure of these particular American Indian individuals was not the same as immediate neighbors. This finding implies that these interbred with a people who had a history of tuberculosis exposure—such as Europeans.

I came to believe that these American aborigines, the Narragansett tribe, were the progeny of Vikings who genetically transferred their exposure to tuberculosis from Europe.

I welcome you now to sail down the same route of investigations that I have sailed. You may draw your own conclusions and feel yourself at complete liberty to express your reservations or agreement.

Timeline of Significant Events

489	St. Brendan, born in Ireland, founded monastery at Confert, Galway in Ireland. He sailed with 17 monks evangelizing across the North Atlantic for seven years.
793	Vikings attacked Linisfarne monastery on a holy island off Northumberland in England, and killed or enslaved monks, nuns, and others.
795	Vikings raided Ireland spurred by shortages of arable land in Scandinavia. Culdees (Irish monks) fled the Vikings and settled in Iceland.
798	Vikings attacked France.
820	Vikings attacked Flanders.
834?	Oseberg Viking longship was found in Vestfold, Norway, now on display in Oslo. It was built around 813 and buried, according to Viking tradition, to honor an important woman who was interred within it.
840	Viking settlers founded Dublin, Ireland.
850	Three Viking brothers were asked to run Russia from Novgorod, Izborsk and White Russia.
859	Vikings raided Mediterranean coast.
860	Vikings attacked Constantinople (Istanbul).
862	Novgorod in Russia was founded by Rus Vikings.
863	Gardar Svavarsson the Dane, of Swedish origin, discovered Iceland.

866	Danish Vikings took York and established a kingdom there in England.
870	Norse Vikings settled on Iceland.
874	Viking longship fleets went to Reykjavik, Iceland.
879	Kiev was established as the center of the Rus Vikings.
886	England was divided under the Danelaw pact and Danish Vikings controlled the northern half, English controlled the southern half.
890	Gokstad ship was built.
917	Vikings defeated Dublin, Ireland.
922	Arab historian Ahmed Ibn Fadlan saw Vikings arrive in Bulgar at the Volga.
930	First true democracy (Alltinget) was founded at Thingvellir, Iceland, by Vikings.
941	Rus Vikings attacked Constantinople (Istanbul) again.
943	Arab historian Ibn Miskaweich described Vikings on Bedaa south of Baku in Azerbaijan.
974	Leif Erickson was born to Erick the Red in Iceland.
976?	Erick the Red Thorvaldson discovered/named Greenland after being outlawed for manslaughter in Iceland and sentenced to three years exile.
984	Scandinavian immigrants occupied Greenland's Eastern and Western settlements.
986	Erick the Red moved to southern Greenland with his family. Led first 28 boats of settlers (14 made it) from Iceland, settled in Brattahlid. We do not know whether any of the other 14 boats were blown off course on beyond Greenland. Bjarni Herjolfson, sailed to visit his father who had just moved to Greenland, was driven west by a storm. He saw America (New Foundland?) but didn't land and returned to Greenland to tell his tale to Erick the Red and others.

991	Viking chief Olav Tryggvason's 93 ships defeated the English at Maldon near Essex.
995	Olav (Tryggvason) I conquered Norway and proclaimed it a Christian kingdom.
1000?	Leif Erickson, son of Erick the Red, returned to Greenland from Norway, bringing first Christian missionary. First Christian church built in North America at Brattahlid. Leif island-hopped exploring coast of North America. He and others sailed to and named Helluland, Markland, and Vinland.
1003	Leif's brother, Thorvald, borrowed his boat and sailed to America. Died the following year leaving widow Gudrid with Leif and his family in Greenland.
1006	Merchant Thorfinn Karlseffni visited Greenland, married widow Gudrid.
1010	Thorfinn Karlseffni sailed from Greenland with Gudrid to found a settlement in North America.
1014	Vikings established L'Anse aux Meadows, Newfoundland, as base camp for exploration and ship repair/maintenance. Artifacts are dated from 990 to 1050 with a mean date of 1014.
1014	Vikings in Ireland were defeated at the Battle of Clontarf.
1026	Icelandic historian Ari Thorgilsson wrote about Irish monks in Iceland in 1026.
1047	Svend Estridsson gained control of Danish throne.
1064	Hardraada gave up Denmark, recognized Estridsson as legal heir to throne.
1066	William, Duke of Normandy, defeated Saxon King Harold at Battle of Hastings.
1067	Ari Frode Thorgilsson was born. Ari the Wise began chronicling genealogies and the history of the Vikings before he died in 1148.

1072/6	German cleric/historian Adam of Bremen wrote on the development of Christianity to the north. After visiting King Estridsson in Denmark, he wrote that where Vinland was discovered, vines grew wild.
1124/6	Greenland became diocese. Episcopal residence placed at Gardar near Brattahlid.
1154	Arab geographer Al-Idrisi described "Great Ireland," Irish territory "beyond Greenland" near Norse Vinland in his atlas.
1200	Saga writing began, committing to paper events passed down orally.
1350	*Book of Knowledge* by Spanish Franciscan described Irish settlement in America.
1350–60	Ivar Bardarsson, Icelandic clerical official, reported the Western Settlement in Greenland was being abandoned. Inuits began to appear near Norse areas.
1380	Greenland-Knarr, trade vessel for Greenland was lost, ending trade between Norway and Greenland.
1387	*Codex Flatoiensis* (manuscripts of the sagas) was completed.
1397	Norwegian, Swedish, and Danish kingdoms merged.
1398	Henry Sinclair had Nicolo Zeno survey Greenland as jumping off place for New World with 12 vessels of monks and fugitive Templars.
1408	Wedding at Hvalsey Church in Greenland. Last written record of Greenland's Norse population.
1448	Pope Nicolas V wrote sad news to bishops describing a calamitous attack on the Greenland settlement in 1418.
1477	Christopher Columbus sailed with the Portuguese to the Azores, Ireland, Iceland and 300 miles north of Iceland.
1492	Christopher Columbus discovered America.

1496/7	John Cabot and his son explored Canada and Newfoundland for England.
1500	Joao Fernandez sailed from the Azores to Greenland and perhaps to Labrador. Gaspar Corte Real sailed for Portugal to Newfoundland.
1501	Gaspar Corte Real sailed again to the New World but was lost. Two of his ships returned. Some say he stayed in America.
1502	Gaspar Corte Real's brother, Miguel, sailed to find his brother but his ship was lost. Some say he stayed in America.
1512	Ponce de Leon discovered Florida.
1519	Hernando Cortes discovered the Aztec Kingdom in Mexico.
1521	Ferdinand Magellan circumnavigated the world.
1524	Giovanni da Verrazzano explored the Northeast American coast. Portuguese Estevan Gomez sailed to Nova Scotia for Spain and then to Bangor, Maine, and Cape Cod. He took (Narragansett?) Indians back for curiosity.
1525	Spanish mapmaker Diego Ribero made charts of the Pacific. Joao Alvares Fagundes of Portugal sailed to Gulf of St. Lawrence and Newfoundland. He set up a colony in Nova Scotia. Luis Vasquez de Ayllon sailed for Spain along the Carolina coast near Wilmington. Some of his survivors returned to Spain after battles with hostile Indians.
1527	The map of Vesconte Maggiolo (Maiollo) depicted "Norman Villa" and "refugio" along the New England coast near Rhode Island.
1527/8	John Rut sailed from England and saw Newfoundland, Cape Breton in Nova Scotia, New England, the West Indies, Santo Domingo, and Puerto Rico.

1529	Diego Ribero's map showed Maryland, New Jersey, New York and Rhode Island as "the land of Estevan Gomez." Verrazzano's brother produced a map of their travels showing "B. del refugio" at Narragansett Bay.
1532	Francisco Pizarro discovered the Empire of the Incas in Peru.
1534/6	Jacques Cartier explored the Great Lakes, St. Lawrence, Newfoundland, Canada, and founded Quebec and Montreal.
1536	Richard Hore of England took men to fish in Newfoundland but short on supplies, resorted to cannibalism before returning to England.
1541	Cartier returned with five French ships to set up a colony near Quebec. He took criminals (including Manon Lescaut) because others feared such a voyage.
1565	St. Augustine, Florida, was founded by Spaniards.
1576/8	Martin Frobisher made three trips to Arctic Canada looking for the Northwest Passage.
1570	Sigurdur Stefansson's Icelandic map showed *Promontorium Winlandiae*.
1577	Sir Francis Drake circumnavigated the world for three years.
1579	Copies of Stefansson's map were made, which are now in museums.
1583	Sir Humfrey Gilbert of England brought 260 to establish a colony in Rhode Island and Connecticut. Bad weather forced them to leave after founding Newfoundland.
1584	Amadas and Barlowe explored North Carolina coast for Sir Walter Raleigh.
1585	Raleigh established Roanoke colony, first English settlement in North America.

1585/7	John Davis sailed three times from Britain to find the Northwest Passage to China but little is known of his findings after he passed Greenland at Davis Strait.
1607	Jamestown was founded.
1607/10	British captain Henry Hudson sought the Northwest Passage but found only Hudson's Strait and Bay before his crew mutinied in 1611.
1609	Capt. Christopher Jones sailed the *Mayflower* to Norway to trade, and possibly to Iceland.
1614	Dutch explorer Adrian Block visited the island off Rhode Island named for him.
1615	A devastating epidemic (smallpox?) decimated 80%+ of Native Americans.
1620	Capt. Christopher Jones brought settlers on the *Mayflower* to Cape Cod, but they soon relocated to Plymouth.
1631	Roger Williams emigrated from England to Massachusetts.
1634	William Blackstone was the first Rhode Island settler.
1636	Roger Williams was banished from Massachusetts for "dangerous ideas" regarding freedom of religion. He settled in and founded Rhode Island.
1637	Pequot-Narragansett War.
1639	The Newport Compact was the basis for the settlement of this area, signed by settlers including early Rhode Island colonial governors.
1643	Roger Williams published *The Key to Indian Language*.
1663	Charles II granted the Charter of Rhode Island & Providence Plantations.
1670	Census of Narragansett Indian population done around Rhode Island.

1675	First decisive battle in King Philip's War was fought against the Narragansetts. Great Swamp War and Massacre involving Jireh Bull's blockhouse.
1676	Burning of Roger Williams' home and others by Narragansetts.
1717	Cyprian Southack developed 8 navigational charts of New England.
1777	Charles Blaskowitz surveyed Narragansett Bay for the British military.

Part I
▼

Vikings

We look deeply at a people viewed universally as primitive barbarians. We find, instead, a peaceable and stable culture of farmers, fishermen, and seamen. Advanced for their day, they bequeathed to the Western world a refined sense of law and formed the first and longest lasting of democratic governments.

Where Do We Start?

Two names cross all cultural barriers and are recognized by nearly everyone worldwide. One is Erick the Red or the Ruddy and the other is his son, Leif Erickson or Leif the Lucky. Vikings were extremely blunt in naming people, as we will learn. Some names I particularly enjoy are Harold Blue Tooth and Sven Rat Nose.

Leif was born about 974 A.D. in Iceland but two years later, his roughhewn, pagan, and combative father, Erick the Red, was exiled. Erick took his family to Greenland. There is much known about Leif such as his words, actions, conversion to Christianity, coloring, height, and modus operandi. His story and his adventure—how it stirs the minds of men!

The mystery of where Leif voyaged has aroused universal curiosity for centuries. His landings have been sought by legions of explorers, scholars, sailors, analysts, and archaeologists. The basic folklore of Iceland has been dissected and examined to a fare-thee-well. Seminars, college courses, replica voyages, full scale expeditions mounted, myriad artifacts studied, and books have been written and filed in libraries worldwide.

I believe that the search can be successfully concluded with the site I found which matches the terrain and the travel time. Perhaps, just perhaps, I have found the place where Leif Erickson and his hardy band of 35 made their epic landfall a thousand years ago. My proposal can be proven or not by those coming along behind me.

Some of the numerous sources on the Vikings include the *Landanama Book*. This may be the most complete record of its kind ever made by any country. It is similar to the English Doomsday Book, but much more interesting and valuable. It contains the names of 3,000 people and 1,400 places. It includes genealogies of the first pioneers and their feats. Ari Frode Thorgilsson, called Ari the Wise or Ari the Learned, who lived from 1067 to 1148, began it. It was extended by Kalstegg, Styrmer, and Thordson, and was completed by Hauk Erlendson, Lagman, and the Governor of Iceland, who died in 1334.

Ari also wrote the *Book of Icelanders*, which goes into detail concerning the discovery and colonization of Greenland, and he mentions Vinland in connection with the genealogy of Thorfinn Karlseffni. (Ari's uncle, Thorkel Gelisson, learned about this from a companion of Erick the Red, says the *Catholic Encyclopedia Online* in an article entitled "The Pre-Columbian Discovery of America.")

We have versions of all the sagas relating to America in the *Codex Flatoiensis*, completed about 1387 or so. This *Codex* was first in the archives of Copenhagen, Denmark, but in 1971 was moved to the Árni Magnússon Institute, an academic institute located in Reykjavik, Iceland. The institute has the task of preserving and studying medieval Icelandic manuscripts containing *Landnama, Heimskringla,* and the Icelandic sagas. These manuscripts were found in a monastery on the island of Flato, hence their name *Codex Flatoiensis* or *Flateybook/Flateybok*.

The saga tales have a ring of plausibility, especially to those conversant with the sea and seafaring life. The word "Viking" has a nautical connotation. "Vik" meant in the Old Norse vernacular a "small bay," possibly "cove." It also identified those people capable of such seamanship, cultural, and military organization that for some 500 years they were nearly invincible. They essentially "owned" the North Atlantic Ocean from above the Arctic Circle to below Gibraltar and even within the Mediterranean. This span in Europe is the equivalent in America as from Greenland to south of Chesapeake Bay.

They were the very first "blue-water seamen," and were alone in that skill for centuries. Around 800 A.D. they settled Iceland, developed a mixed Norse/Gael folk, and overpopulated the usable area within 200 years. Icelanders looking for more land and resources were traveling as far as Constantinople by the time of the Vinland voyages.

About 984 A.D., emigrants settled Greenland, which then became a long-lived and moderately successful European outpost for more than 400 years. It was from that newly founded colony in Greenland that the Vinland voyages originated. It was from Greenland and Iceland that people joined in the adventure to Vinland in America.

What an inspiration they are! From before 1492 until today they have given impulse to generations of people to locate that place where Leif Erickson set foot in the New World. The sagas are so difficult to access that most feel constrained to deal only with the famed Leif and nearly forget his contemporary friend and relative Thorfinn Karlseffni who followed the tracks to Vinland as well. This man, and his wife Gudrid, left a greater number of clues to the location of Vinland than had Leif. In fact, Leifsbudir (Leif's booth or camp) came to be well visited for some 20 or 30 years by Karlseffni and others.

More than 300 persons made the trip, one individual traversing twice and a single ship making passages three or more times. It is from this complexity, the sheer numbers of those who returned and preserved their tales that some of the confusions of the sagas originated. At least five individuals have contributed their stories.

This book deals with the pertinent Vinland sagas because the tales are descriptive enough to have their own "built in" crosscheck for plausibility. A Viking sailor, Bjarne Herjolfson (or Herjulfsson) was blown off course and became lost but saw a new land. Later, Leif purchased Bjarne's ship and sailed to find that new land. When Leif returned to Iceland, his brother, Thorvald, criticized him for not exploring sufficiently. Thorvald took the same boat and returned to Vinland for that purpose.

Another Norseman, Thorfinn Karlseffni, a merchant who married Leif Erickson's sister-in-law, Gudrid, contracted with Leif to use his site. So did Leif's half-sister, Freydis, in later years. Leif had a sort of time-share for relatives. The Vinland settlements are the stories of a small group of people who were joined by others, friends, crewmen, and servants. Their tales hang together and they make up a dramatic and resplendent tale of the sea. Most others and I believe them to be essentially true and believable, and we make them the basis of our research.

Northern Europeans, Teutons, Germanic peoples, Norsemen, and Scandinavians had arisen in the Baltic environs where they developed the subculture called Vikings. Nautical Norsemen developed this subculture by advancing boat building and seamanship, which allowed them to populate the island archipelagos of the Baltic Sea and North Atlantic.

Their remarkable boats and ships made Vikings powerful throughout half of the world. They had the nautical ability to move about Europe, Near East Asia, and America. The "longship" and workaday "knarr" design was well adapted for rowing inside through often windless fjords, then sailing when winds permitted outside. The rudder strikes the modern eye as awkward and peculiar, yet it had a distinct advantage in that it could turn the boat even if forward motion was nil. Other rudders required a current or the thrust of a propeller. This rudder, mounted on a swivel, could operate by a sort of rowing motion called "sculling," but the thrust can be sideways rather than rearward as sculling is normally used.

A tent-like affair aft of the mast was the captain's home afloat, where he and his closest companions resided,. This is where valued possessions and perhaps food was stored and defended, a "ship's keep" of sorts. It is this author's opinion that it was called a "budir" or booth because it was set up with varied types of walls. Leif was said to have moved his budir ashore at his landing, conveying the same idea Americans do when they "put down roots" or "pull up stakes," hence, "Leifsbudir, my camp here."

Unseen just forward of the mast were casks used for water storage. Rainwater could be drained from the sail here and this system prevails today even to larger sailing vessels always in front of the mainmast (the "scuttlebutt" or social center) of the ship. Forward of the scuttlebutt would have been crew and passenger territory, with a few sheep or goats, which were useful for milk and meat. Domestic fowl, sheep, goats, swine, and cattle were all taken on long voyages. Cattle were known to have been to Vinland on a very lengthy trip, including at least one bull mentioned in the sagas.

Live creatures could aid navigation. Goats were said to be able to detect land by smell long before a lookout could see it. In fact, Iceland was said to have been discovered by the use of captive ravens by an explorer appropriately known as "Floki Raven."

Rope is the sailor's constant worry, interest, and labor. Vikings were said to have used braided animal skins with the result that anchoring the ship was riskier than beaching. Even beaching is not as simple as it might appear. The periods and ranges of tides and phases of the moon all must be considered when landing. Neglect of this can be fatal such as when Thorvald Erickson could not launch at Crossannes. Oars may have served not only to row boats but also to keep them upright at landings by being fastened vertically at oar ports outboard.

Such a ship would seem pretty crowded with 35 men on Leif's crew or 30 with Thorvald. Along side of the ship were items such as seamen's chests upon which crewmembers might sit to row. Everyone rowed on such boats and that helped to keep them warm. Fires from cooking might have helped as well but little of that occurred on ships. We are unsure exactly where the cooking was done but smoke and aromas had to be considered in the placement. Cooking would not have occurred every day because dried and salt meat, fish, berries, and cheese were known staples.

At the head of the ship, a latrine then as now would have been present. The odors would have constantly been blown away. A simple bucket may have been all that was used.

Using these remarkable ships, the Vikings peopled Iceland near 800 A.D. and Greenland near 984 A.D. From fjord travel to island-hopping to the "blue-water ocean" was a natural progression. The "deep" was theirs and theirs alone worldwide for more than 500 years. They established a shipboard culture that is still maintained aboard sailing vessels.

Vikings have been considered barbaric and primitive thugs. This impression is one of the major factors limiting serious study of Scandinavian history and Vinland. In fact, Norse Vikings were a peaceable and advanced population of medieval Europe. Their style of living was mainly as small groups of farmers and fishermen making obeisance to what we call "petty" princes. They did not have major cities or political centers. Universally accepted cultural mores and customs rather than kings and priests governed them. They looked, acted, and dressed like other Europeans.

They probably got their bad reputation because those were primitive times. The Vikings farther east became a powerful force in trade along the coasts of Europe with influences as far away as Turkey and North Africa. Some became more avaricious when silver and gold gained increasing meaning within the culture.

When men and ships are remote from law and observation, many evils can occur. Small fleets under ambitious captains sometimes made it their habit to land on foreign shores as outright pirates. Later, some of these landings became so well organized that collections of tribute were sometimes called "taxes." Those "taxes" resulted in immense quantities of silver bullion to Scandinavian kings. The word "taxes" has ultra-sensitive connotations to all Englishmen.

Viking targets were often churches and monasteries in England and Ireland, whose religious devotion made them even more sensitive. Norse genes are strong in the Irish body, yet the word "Viking" is still used there as a pejorative equaling "bogeyman."

Viking aggression lessened after Norsemen became Christianized. Gradually, Vikings withdrew, leaving their history to be written by their enemies.

A Brief History of Vikings

I will give only a summary of Viking exploits so that their exploration and sailing abilities can be appreciated.

The Vikings raided and traded around a large part of the world from the 8th to the 11th centuries. The Vikings from Norway and Denmark (Danes) generally went hunting for new land in the west, southwest, and in established colonies.

They colonized the coasts and rivers of Europe; the islands of Shetland and Orkney; Scotland; Ireland; England; Normandy; the Faroe Islands; Iceland; Greenland; and Newfoundland, and reached even further south in the Americas. York, England, was taken in 866 A.D. and became the Danish capital in England. The Norse kings of Dublin were at war with the Irish kings during the 9th century. Resistance to the Norse kings built steam until the final confrontation in 1014 with an Irish victory, but Norse descendants and influence continued for many years.

In 874 A.D., Viking longships went to Reykjavik, Iceland. They brought a Norse chieftain, his family, livestock, and established a settlement to fish and raise sheep. Fish were not only an important food but fish oil was used for lamps, especially cod (cod liver oil). Interestingly, the fish bone protein, collagen, is used for radiocarbon dating of human burials and other sites of human occupation.

Two centuries after they landed there, the estimated population of Iceland was around 75,000 people. Meanwhile, they had formed the first European settlements on the American continent, naming them Greenland, Markland, Helluland, and Vinland. In fact, there is an old saying: "Vikings conquered the world one island at a time."

The Vikings from the area of present-day Sweden usually went east and southeast, and engaged more in trading and stealing valuables. The Swedish Vikings traveled as far as France where they attacked Paris, Sicily, Jerusalem, the Caspian Sea, and Baghdad, as well as Constantinople. A monk named Nestor, who lived in Kiev, wrote about how they settled in Russia. First, they levied taxes on the Slavs until they were routed out. However, when anarchy overwhelmed the country, they sent for Vikings to be kings around 850 A.D. Three brothers responded and ruled from Novgorod, White Russia, and Izborsk. Their trade networks extended as far as China.

The Vikings, always the minority wherever they went, always became quickly assimilated with the natives in their settlements. Joskim Hansson of Lulea University in Sweden wrote about an Arab writer named Ibn Khordo Adbeh who described the Swedish Vikings he encountered around Volga Bulgaria in the early 900s.

> *These merchantmen speak Arabic, Persian, French, Spanish, Roman, and Slavic. They travel from the Occident to the Orient. From the Occident they bring with them eunuchs, female slaves, little boys, fabric, skins of different kinds, and swords.*

In 922 A.D., the Arabic messenger Ahmed Ibn Fadlan saw the Vikings arrive in Bulgar, at the bend of the Volga. James Montgomery wrote about Fadlan's book called the *Risala*. Fadlan was disgusted yet fascinated by Vikings and wrote this:

> *I have never before seen such perfect bodies ... They are the filthiest of Allah's creatures ... Each morning the girl comes early in the morning with a deep dish of water. She gives this to her master who in turn washes his hands, face, and hair ... blows his nose and spits into the bowl ... the dish wanders from man to man until everyone has washed himself in the water ... They have sexual intercourse with their slaves while their friends are watching.*

Abdeh also described a Viking burial. The body was finely clothed, placed on cushions in a sitting position in a small tent, which was built in the middle of the boat. He had several items, which could be useful on his way to the land of the dead, such as a harp, food, axes, a sacrificed dog, two oxen, two horses, and one hen. One of the man's female slaves was chosen to follow him. She was intoxicated with alcohol, brought to the chief by six men, each of whom had intercourse with her before an old woman killed her with a knife while the men were strangling her with a rope. Then the relatives set the ship on fire, and then put the name of the dead man and the name of their king on a wooden pole erected in the ashes.

There is a book by Dr. Michael Crichton called *Eaters of the Dead* which purports to be about the travels and observations of Ahmed Ibn Fadlan. The first three chapters appear to be properly researched but readers should understand that the rest of the book involves the stories from *Beowulf*. As Crichton says, the book was written "based on a bet" that he could make *Beowulf* readable. This Harvard Medical School graduate not only did that, but *Eaters of the Dead* was made into a movie—*The Thirteenth Warrior*, with Antonio Banderas and Omar Sharif.

The Arabian historian, Ibn Miskaweich, wrote in the year 943 A.D. about the people who came with axes, swords, long knives, spears, and shields. Joskim Hansson of Lulea University is again the source of this information. A custom that arose from their extensive travels was to leave behind any number of men and parties for one reason or another. Viking settlements and genes are common in the British Isles, France, Iceland, Greenland, Newfoundland, and somewhat common in Russia, Turkey, and the Mediterranean.

The image of Vikings described by Muslims (who are commanded to purify the body five times a day before entering mosques) sounds rather harsh. Let us add some additional descriptions so that they may be imagined more accurately in Norway, Iceland, and Greenland.

Vikings were not giants, and averaged five feet six inches, some three inches taller than Englishmen were. They were often clean-shaven, however many chieftains were bearded. They dressed more or less like other Europeans but with their own decorative style. The life of a Scandinavian was mostly spent in fire-lit gloom among family and friends for months on end. Dealing with others in close quarters for long periods requires a certain mentality and code of behavior peculiar to most people of the North. Despite my hearing impairment, I have observed that Scandinavians seem to be the only people who can manage to shout in a whisper.

Males wore a tunic of hip length similar to a Russian "peasant" dress. They omitted the tunic when temperatures permitted and were naked to the waist when possible. Trousers were sometimes close fitting. At other times, they wore baggy pants like the Dutch knickerbockers. They had a stylized short cape clasped at the right shoulder leaving the right arm handy for drawing a sword. Headgear was often a close-fitting leather cap such as the Tollund man and caps such as those in the paintings of Flemish Peter Bruegel (in The Netherlanders

Proverbs and The Adoration of the Kings.) When helmets were worn, I think there must have been a helmet liner to protect the brow from frigid steel. They wore wool socks and loose leather shoes similar to Indian moccasins. The helmet never had steer horns mounted. Those would have been too easy for an opponent to grab. Crafted furs were worn but seem to have been rare and expensive.

Horses (a pony size used for meat, milk, and as pack animals) were a common factor in Scandinavian life. So were cattle and hogs. Their cattle were rather small and easily transported by boat. In English, the small cattle were called "Neat cattle." Their hooves were the source of "neat's-foot oil" for preservation for leather.

Women wore ankle-length dresses and a longer cape clasped at the neck with a characteristic double clasp, an apron, and a kerchief over the hair. Labor was their usual lot and they may even have had slaves or servants from Ireland, England, and numerous other countries. Norwegian women had a peculiar method of milking goats, which relied upon confidence in the good nature of the goat. They straddled the goat facing the rear. Some say that the view from there was not too good but did the job. Still, the goat horns in close proximity to vulnerable anatomy seems unusually brave, but perhaps not to a Viking. To each his own.

In each household, one dominant woman held the keys to the multiple storage chests in the household on a ring hung from her belt. For social position, or at the death of her spouse, she might be the actual "Eirick" or chief of the homestead.

Viking chiefs had their own style. Some historians describe them as being like "dandies." It may come as a surprise that they wore "scents" and used eye coloring. Since they had servants, they could afford to dress well and "put on airs." While lower class men might be armed with axes, pikes, or spears, the Eirick (Earl, Jarl, or Chief) would always have a typical sword. Double-edged and long, it was a lethal weapon in the hands of a strong man. In one duel, it was said that a participant had nine shields shredded to pieces. When the tenth failed, the sword immediately penetrated the skull of the harried and unlucky defender, who doubtless had little time to consider his intemperate judgment in trial by combat with such a powerful adversary.

Shields were more commonly round, but a few seem to have been of "lozenge" shape. Chieftains helmets had cheek guards similar to ancient Greek helmets. They were highly valued, decorated with plumes and effigies, and concealed the whole face. The full beard hid what was not protected. This menacing look is where we come closest to the reputation of the fearful Vikings.

In Iceland and Greenland, their form of democracy was feudalism where "Earls" ("Jarls") achieved status depending on the size of their property, number of ships, and the quality and amount of followers. Within the bounds of each "stead" or "vik," the social order aimed for stability, efficiency, and prosperity. For, after all, should the crops fail or food supply be interrupted, there was only starvation to face for many months.

The Icelandic and Greenlandic sagas derived from passing valuable historical information mostly on long winter nights in the North by telling tales. Some days have little light at all in the far north. Catholic monks set many of those sagas down in Latin as historical records. Some were written in "runic" which method passed from the scene. The majority of the people could not read or write, even if they were intelligent. The word "saga" is Old Norse for "sayings." Thus, the sagas were educational, dramatic, and were told as narrative stories with poems to inspire easy memorization. Modern teachers can appreciate how important those qualities are to transfer information that will be retained.

Story telling is as old as human civilization. Story telling was our oldest form of literacy. Epic stories were present in most cultures and were generally about heroes and heroines, gods and goddesses and battles. They were told in the form of song, drama, poems, allegories, or parables in almost every culture that we know of.

In the Nordic countries, story telling was a valued tradition. Across Iceland, Norway, Sweden, and even Finland, these myths would be told and retold until they were etched into human memory and known even until now.

John Pfeiffer wrote about how the ancients remembered things in *The Creative Explosion: An Inquiry into the Origins of Art and Religion* on p. 223.

> *Some of the poets, scholars, and orators of classical times were accomplished mnemonists. They spent hours training themselves to reconstruct small worlds in their heads, complexes of many buildings with thousands of memory-places in each building. One Roman teacher of rhetoric would listen while each of the 200 or more students in his class recited successive lines of a poem, and then recite the entire sequence backward ...*

Interested individuals collected the Vinland sagas as the population became more literate. Therefore, they are recorded in various styles and forms. We don't know that they are all true but many sound plausible—and they are all we have to work with.

Viking Navigation

The sagas do not describe Viking navigation in much detail, probably because it was presumed that everyone already knew about it. Scandinavian accomplishments in shipbuilding are well known. They first used oars, and then added a keel. The keel gave strength and a broad heavy base from which a mast could be supported for sails. By great artisanship, they planed and caulked usually oak wood to build sleek speedy boats. This speed enabled them to raid and retreat before troops could be assembled to fend them off. Their speed, even laden with over 30 tons of cargo, was probably as much as 18 knots in a "surge," possibly averaging 8–10 knots for a possible 200 miles a day in favorable circumstances.

But how did they aim for destinations thousands of miles away? It was one thing for a ship to be blown off course and to land upon an unintended place. But how could the Vikings find an area like Vinland and sail back to it several times unerringly? In at least one case, they used a rather quick route, which means they may not have hugged the coastline. How did they do it without magnetic compasses, astrolabes, or other devices? Let me give a few theories by those who have studied Viking navigation.

"The Viking" by Howard La Fey was written for the *National Geographic Society* in 1972, and described how the Vikings had decades of carefully won practical knowledge of the positions of the sun and the stars. Of course, in northern climes, celestial navigation could not be used a lot of the time. Much of the year, it was either all light or all dark. Vikings usually tended to stay close to land but when they risked the open sea, they had to know how far north or south they were from home.

Fey told of how random scratches on a Greenland "bearing dial" were found to actually mark the shadow of the sun on a particular latitude, enabling one to find the directions at times other than high noon. The bearing dial helped determine the position of the sun and moon when the Pole Star was not visible. The position of the Pole Star or Polaris, (two stars in *Ursa Major* or Big Dipper point to it) could aid them as well. Located in the night sky, directly over the north celestial pole, the distance from the North Star to the horizon was compared to the height of the star when they were at home. Thanks to a wobble in the earth created by the gravity of the sun and the moon, the star most nearly at our north celestial pole keeps changing. A thousand years ago and now, that star is Polaris (only one degree off true north) but 14,000 years from now, it will be Vega.

No matter where one is, Polaris is always north and it appears as if all other stars rotate around it. This measurement helped mariners determine their latitude, and everyone observant to know from the angular position of the stars. It was a most perfect timepiece, both what time it was and how long to the awaited sunrise. They may have also used a notched stick or the mast of the ship to look past the star and note how far up on the stick the star appeared. An experienced pilot could see that he was at the same latitude if the star was seen against the same mark. A higher notch meant the ship was at a higher latitude, or nearer the North Pole. The Vikings produced latitude tables for certain stars including the sun.

Another method for navigating was to observe migrating animals. The experienced sailor would use sightings of whales known to be half a day's sail south of Iceland or migrating birds such as geese to help locate land.

An article entitled "Secrets of Ancient Navigation" by Peter Tyson published by NOVA online described how Norsemen relied on the behavior of birds. A sailor might observe an auk with a full beak and conclude it's heading toward its rookery; if empty, it's heading out to sea to fill that beak. One of the first Norwegian sailors to Iceland was Raven-Floki who kept ravens aboard his vessel. When he thought he was near land, he released the ravens, which he had deliberately starved. Often as not, they flew directly toward land, which Raven-Floki would reach simply by following their lead.

Things could go wrong for even the most experienced pilot, however, and strong storms could blow Viking ships off course. Franck Pettersen of Tromsø, Norway, wrote a paper called "The Viking Sun Compass or How the Vikings Found their Way Back from New York 1000 Years Ago." He described how the Vikings were the first known people to use the keel, which was necessary to keep a stable course when they crossed the oceans. Their sailing route was between the 61st and 62nd degree north on a due western course from Norway to Greenland.

He theorized about how they managed to keep their straight course across this great distance when they did not sail along coasts. One of the theories is that they used Icelandic feldspar to find the general direction. The feldspar sunstone was used on the days when fog or clouds obscured the sun. The stone, a mineral called Icelandic spar (optical calcite), would change color as it was turned in the light. A certain color would indicate the position of the sun through the fog or clouds but could only be used when there was at least a sliver of blue sky. If the stone is held in the fog or under overcast skies, it will appear dark (as it does not allow diffuse light to pass through). However, if pointed directly at the sun, even in fog, it allows the direct rays of the sun to pass through, so that it will light up.

Consequently, the Northmen moved this stone around in the fog. When it lit up, they knew that it was pointing directly at the sun. From there, they could figure their way. However, feldspar had an error of plus or minus 30 degrees from the set course. Spar stones (calcite crystals) have been found in some Viking graves so this may have been more common than we realize.

Several Viking navigation sites mention that Hrafn's saga written around 1230 says: "the weather was thick and stormy ... The king looked about and saw no blue sky ... then the king took the sunstone and held it up, and then he saw where [the Sun] beamed from the stone." However, that saga does not say that the stone was used for navigation. That citation is said to be found in *Hrafn's Saga Sveinbjarnarsonar*, edited by Gudrun P. Helgadottir, published by Clarendon Press in Oxford in 1987.

Pettersen concluded his article with these words, quoted from the *Planetarian*, Vol. 22, No. 1, March 1993, used with permission:

> *Vikings before Leif Erickson went from Norway through Gibraltar and all the way to the eastern Mediterranean. Vikings had crossed the Atlantic Ocean and settled at Iceland and Greenland.... Having reached Newfoundland, there was a pressure on the next Viking captain or Viking chief to go further south to explore new lands to prove that he was a man, and that he was a skilled navigator and explorer.*

A Viking sun compass found in Greenland had different hyperbolas or (gnomon) curves, and the north direction was clearly marked with 16 small cuts crossing a long line. Counting the spikes from north to the right, you have 90° or due west at spike number 8. This also indicates that the Vikings divided the compass into 32 directions before the magnetic compass was used in Europe. The sun compass draws on the fact that the sun's shadow from the tip in the middle of a disk describes different hyperbolas at different times of the year. They would rotate the disk until the shadow of the tip falls on the hyperbola, and the general directions are given with an accuracy of a few degrees.

Several Viking navigation writers believe that a sun shadow board was used. It is similar to the sun compass. It was a circular board with a tip in the middle, and the board was allowed to float in a bucket of water. Since most assume that the Vikings sailed across the seas in straight lines using a particular latitude, this method would help them keep to their desired latitude. However, it would require sunlight to cast a shadow.

While not pertaining to navigation, one might wonder how several ships kept within distance of each other. Samuel Morison reported that during fogs or night they might have drummers and singers whose sounds kept ships aware of where they were.

By the time of Columbus, Viking sailing methods had been adopted by many European countries. There is no doubt that they set the precedent for ships that were to become the great invaders from Europe to the Americas.

Even Columbus was versed in Viking sailing vessels and methods before his adventures to the lower Americas. In 1477, Columbus shipped out in a Portuguese vessel (perhaps a member of the Corte-Real expedition) exchanging wool, dried fish, and wine between Iceland, Ireland, the Azores, and Lisbon. Samuel Morison wrote:

> *The master of the vessel in which Columbus sailed in February 1477 went exploring to the north of Iceland for a hundred leagues before returning to Portugal.*

A league is three miles and it is possible that Columbus' ship made landfall 300 miles northwest of Iceland on Greenland where earlier Vikings had settled. Columbus read many books, made notes in them and on maps regarding the travels of Marco Polo. He was probably familiar with the travels of the Vikings and others. Columbus, like us, may have wondered exactly where the Vikings had found Vinland.

"We might consider as well, that the location and latitude of Rome must have had salient interest to these early and ardent Catholic Christians. The issue of wine in Vinland became one of religious aspect as Catholic doctrine specified grape wine, and only grape wine, could be used at Mass. We can well imagine that both latitude and location of Rome were important considerations to the Vinland Voyagers. This is a safe assumption because one of them, Gudrid Thorbjornsdottir, actually visited Rome in later years."

L'Anse aux Meadows: A Proven Norse Settlement in America

Those familiar with this subject will, of course, be reminded of the discoveries and developments of the Norse Site at L'Anse aux Meadows on Newfoundland in Canada. While many feel that evidence is sparse, some believe that this site must, by default, be the one that Leif Erickson discovered. Certainly, it is Norse and its discoverer, Dr. Helge Ingstad, pursued the theme that it might well be the site of legend. But over the years it has come under some suspicion for Leif's encampment because it bears little resemblance to what the sagas say about both it and its relationships with varied other Vinland sites.

Helge Ingstad had studied Sigurdur Stefansson's Icelandic map of 1570, which identified a place that looked like the north coast of Newfoundland as "Promontorium Winlandiae." Stefansson was a schoolmaster in Skalholt, Iceland, and his original map has not survived but copies were made. A copy in 1579 showed Norway on the right, with Ireland and Britain in the bottom right hand side. It showed Iceland at the top with a glacial sea beneath. Greenland was to the left, Helluland and Markland below Greenland, also on the left, and finally the land of the Skraelings (screechers or foreigners) bottom left. The "Promontorium Winlandiae" is protruding. The islands of Faroe, Shetland, Orkney, and "Frisland" (a confusion of some cartographers with the Faroe Islands) are also shown in the right center area.

Stephansson probably based his map on traditional knowledge preserved from the time of the Vinland journeys. The Icelandic map of 1579 was intended to be a copy of Sigurdur Stephansson's 1570 map of Iceland. The 1579 copy is in the Royal Library in Copenhagen.

Journalist Helge Ingstad and his archaeologist wife told how they arrived at the small fishing village in 1960 and asked a fisherman if there were any strange ruins in the vicinity. The fisherman took him west of the village to a beautiful place with lots of grass and a small creek and some mounds in the tall grass. There were remains of sod walls. Fishermen assumed it was an old Indian site but Ingstad knew that Indians didn't use sod houses. For the next eight summers, Ingstad and his wife and an international team of archaeologists excavated the site.

Their first reports of discovery were not believed. They came upon remains of a blacksmith shop with a huge flat stone for the anvil, with charcoal and lumps of iron scattered about. A few of the pieces had been forged into nails. This was the earliest evidence of iron processing in North America.

The Ingstads uncovered the outlines of eight houses, three of which were where the people lived. Built in the style of Icelandic houses, the walls were six feet thick, two layers of sod between a layer of gravel for drainage. The roofs were made of turf laid over a timber frame. Radiocarbon analysis dated the artifacts at between 990 and 1050, which was the time of Erickson's and subsequent expeditions.

The site had been used mainly to work on boats, build them, repair them, and maintain them. However, it included an interesting find. Saunas were used in Scandinavia and here. The bathhouse contained a large number of brittle burned stones as a sauna, the earliest known in North America according to Peter James *et al.* in *Ancient Inventions*.

In later excavations, Dr. Birgitta Wallace of Parks Canada uncovered even more artifacts confirming the site's Viking origins. Geochemical analyses of pieces of jasper, used to make sparks for starting fires, revealed trace elements found only in Greenland or Iceland. In the ground outside one of the houses was a bronze pin with a ring head, in a Norse style and probably made in Britain. The Vikings used such pins as fasteners for their cloaks.

Among the 800 artifacts, archaeologists also found soapstone oil lamps, a bone needle and more iron nails. Some of the smaller houses appeared to be workshops for carpenters and weavers. A spindle whorl attested to work with textiles, and since Vikings considered this women's work, at least some of the expeditions must have included women.

The absence of evidence of any barns, Dr. Wallace said, indicated that this was not a farming settlement, but a base camp for the Viking sailors as they surveyed the region for trade goods and likely places for more permanent occupation.

Long after the Vinland settlement, Vikings from Greenland repeatedly visited the shores of Labrador for timber, supplies, and food. Archaeologists have found Norse artifacts, including spun yarn, there and on northern Baffin Island. Neither the early Eskimos nor their immediate Inuit successors spun yarn or worked wood by sawing, nailing, and mortising. Therefore, Dr. Patricia Sutherland of the Archaeological Survey of Canada said that the artifacts pointed to extensive contacts with the Greenland Norse for several centuries. A Norse penny minted in the late 11th century turned up at an Indian site in Maine, but many think it got there by trade. It should be noted, however, that the bushy Maine Coon Cat, the official Maine State cat, is thought by most breeders to have been introduced to America by Vikings because its closest relative is a Norwegian breed of cat.

Most now concede that L'Anse aux Meadows was not Leif's Vinland home site. It seems instead to have been a waystop, campsite, or boatyard for short stays. One reason for this is the finding of butternut shells there, demonstrating that the land of grapes was much further south. The Vikings ventured back and forth between L'Anse aux Meadows and grape-growing areas bringing back the butternut shells, which grow in the same area as grapes.

While scholars accept that perhaps Leif landed there, many believe that better possibilities might be found further south. In the sagas, a confusion arises with the expedition of Karlseffni who is generally felt to have traveled as far south as New England or even further. Therefore, it seemed sensible to explore southerly in search for Hop. Moreover, while we are about it, we might as well keep our attention fixed on Leifsbudir, as well.

However, the archaeological discovery at L'Anse aux Meadows is the only settlement that shows definite Norse presence in North America. A total of 148 Carbon 14 dates from the site, which was covered over with more recent Indian or Eskimo artifacts, range approximately from 990 to 1050 with a mean date of 1014. The site was used for only a few years at most. The site accommodated 70–90 men and women from both Greenland and Iceland.

The L'Anse aux Meadows site contains no evidence that grapes ever grew there. However, butternut seeds that grow near grapes were found so these Vikings ventured into grape-growing areas. Why are grapes important? Vikings loved to return to their homeland with goods they could show off to impress others. Wine was such a prized possession that Vinland was the perfect name to draw people, just as Greenland had been chosen to draw people to that unwelcome frozen island.

Adam of Bremen wrote about 1075 A.D. that vines grew at Vinland. Adam's data was independent of the sagas, but may have come from the same oral tradition. A German cleric and historian, he was interested in places where the Christian religion had recently been introduced. He wrote a book in Latin *(Descriptio Insularum Aquilonis*, 1072–1076) on the development of Christianity in the North. To gain information, he visited Danish king Estridson who was interested in history. On his return to Bremen, Adam devoted the fourth volume of his book to a description of Denmark, Sweden, Norway, and Iceland. He made the following comments:

> *Moreover, he (Svend Estridson, King of Denmark from 1047–1076) spoke of an island in that ocean (northern seas containing Iceland, Greenland and Vinland) discovered by many, which is called Vinland, for the reason that vines grow wild there, which yield the best of wine. Moreover, that grain unsown grows there abundantly is not a fabulous fancy, but, from the accounts of the Danes, we know to be a fact.*

In their introduction to *The Vinland Sagas*, Magnusson *et al* concluded that L'Anse aux Meadows was probably not Vinland but it does prove the presence of Vikings in North America.

The transcript of a NOVA special on 5/5/2000 about the Vikings and L'Anse aux Meadows may be of some interest and is included here. Helge Ingstad, who discovered the site, used the sagas for directions, much as I have done. In fact, he began his explorations a mere seven miles from Pettaquamscutt. He, too, thought it looked like a prime candidate but he went on to find the Newfoundland site. I learned this through personal correspondence with the wonderful gentleman, Helge Ingstad, and through his original book.

Also on the NOVA broadcast was Birgitta Wallace, who became Director of the Parks Canada Project. The narrator described how Ingstad traveled up and down the Newfoundland coast, searching for some sign of a Viking presence. Finally, in 1960 a fisherman from the tiny village of L'Anse aux Meadows led him to the ruins of what may have been houses, outside of town.

Ingstad said,

> *I didn't know of course who had built these houses. It might have been Indians, it might have been Eskimos, it might have been fishermen, it might have been whal-*

ers, and at last, it might have been Norsemen. But only excavation could tell the story.

    Ingstad's wife, archeologist Anne Stine, led the excavations to uncover the true identity of the ruins. Slowly, the foundations of turf-walled houses came to light, along with ancient fireplaces. They looked like Viking buildings, but the design alone wasn't proof. They needed some artifacts that were indisputably Viking. Finally, they found what they were looking for: a ship rivet made of iron, a bronze cloak pin, and a spindle whorl—exactly like those found in Iceland. The objects were clearly Viking in origin, not Native American. Today, the settlement at L'Anse aux Meadows has been reconstructed with solid sod houses, just like the homes in Greenland and Iceland. Although Ingstad felt the location, right next to the sea, was a natural for a Norse site, archeologists like Birgitta Wallace realized this wasn't a typical Viking farm.

    Wallace said on the NOVA program,

> *One reason we know why it's a base camp and not a normal Norse farm is its location ... right on the coast, completely exposed to the sea winds. In Greenland, all the sites, and even in Iceland, are located at the heads of fjords, where it is nice and protected ... all the buildings on the site are a very typical Icelandic or early Greenlandic style. But the way they are combined is totally different ... it was not a permanent farm—not an attempt to establish a colony here.*

THE END OF THE GREENLAND VIKINGS

We can learn from the Vatican Library, archaeological references, and other sources when the Vikings came to an end. From Rasmus Anderson's *Norroena Society* work of 15 volumes, a letter from Pope Nicolas V to the Bishops of Shaoltensus and Olensus shows that the end of the Vikings came by the time this letter was dated September 20, 1448. Here is an excerpt:

> *We have heard with sad and anxious heart the doleful story of that same island [Greenland], whose inhabitants and natives, for almost six hundred years, have kept the Faith of Christ, received under the preaching of their glorious evangelist, the blessed King Olaf ... the barbarians ... attacked this entire people in a cruel invasion, devastating their fatherland and sacred temples by fire and sword, leaving in the island only nine parochial churches ...*

Jared Diamond has described why Vinland and Greenland collapsed in his remarkable book, *Collapse: How Societies Choose to Fail or Succeed*. His synopsis of the reasons for failure of Vinland was:

> *In short, the Vinland colony failed because the Greenland colony itself was too small and poor in timber and iron to support it, too far from both Europe and from Vinland, owned too few oceangoing ships, and could not finance big fleets of exploration ... It is no surprise, then, that 500 Greenlanders ... could not succeed at conquering and colonizing North America.*

Diamond also speculated about why Greenland failed. He asserted that it was because of climate change, a decline in friendly contact with Norway, increase in hostile contact with the Inuit, and the conservative outlook of the Norse.

Actually, not all historians and archaeologists are convinced that there was an increase in hostile contact between the Inuits and the Norse. Brian Fagan stated in *The Great Journey* that the arctic people like Inuits and the Norse, needing to live in close proximity for many months, developed social patterns that avoided conflict.

Diamond's arguments include the fact the Norse considered themselves farmers and allowed their livestock to deforest the land, without noting that the Inuit made up their main diet from seafood.

The story of the Viking colonization in the New World has endured for over 960 years through interpretations from folklore, memory, runic script, Latin, and Icelandic to English. Through all that time, from all that distance, over so many peoples' tongues from that fated settlement of human daring, the actual vision of happenings has been transmitted to us.

The lost ruins of the old Viking settlements on Greenland became known sometime in the 16th century. Interest was revived as a host of individuals (perhaps even the great Admiral Columbus) made strenuous efforts to find the precise locale of Leifsbudir where the son of Erick the Red had built his houses.

It may be of interest to know more details about archaeological findings of Erick and Leif excavated in the early 1990s in Greenland. "The Fate of Greenland's Vikings" by Dale Mackenzie Brown was published February 28, 2000, online from *Archaeology Magazine,* the organ of the Archaeological Institute of America.

Brown reported on the ruins of a five-room stone and turf house from the mid-fourteenth century, excavated in the early 1990s. It should be mentioned that when wood was scarce, turf was used just as western American Indians used sod when wood was unavailable.

Brown talked with archaeologists Jette Arneborg of the Danish National Museum, Joel Berglund of the Greenland National Museum, and Claus Andreasen of Greenland University who excavated a Viking colony on Greenland's grassy southwestern coast.

Sagas and accounts had told how 24 boats of settlers set out from Iceland in 986 to colonize Greenland, the island explored earlier by Erick the Red, but only 14 made it. More soon followed. Under the leadership of Erick, the colonists established dairy and sheep farms, built churches, a monastery, a nunnery, and a cathedral with an imported bronze bell and green tinted glass windows.

Around 400 stone ruins have suggested to archaeologists that the population may have risen to about 5,000. The Greenlanders traded live falcons, polar bear skins, narwhal tusks, and walrus ivory and hides with Norway for timber, iron, tools, raisins, nuts, and wine.

Excavations at Erick's farm in Brattahlid (which means "Steep Slope") revealed the remains of a church, originally surrounded by a turf wall to keep farm animals out, and a great hall where settlers cooked in fire pits, ate their meals, perhaps recited sagas, and played board games. Behind the church, they found ruins of a cow barn, with partitions between the stalls, including the shoulder blade of a whale in this. Archaeologists speculated that as Icelanders ran out of trees and wood, they resorted to the use of whalebones or other things to equip their dwellings and structures.

In 1961, workers discovered near the barn a tiny horseshoe-shaped room. Even though the church had been already found, archaeologists assumed it was a tiny chapel built for Erick's wife, Thjothild, because burials were next to it. They concluded this was because they found three skeletons close to it.

Medieval church records show that those buried closest to the church were first in line for Judgment Day. Thus, the archaeologists guessed that they were Thjothild, Erik and their son, Leif, who around the year 1,000 sailed from Brattahlid to America. Rather than being buried by the main church, they guessed this might have been a private chapel for Thjothild. She had converted to Christianity when her son, Leif, came back to tell that he had been commissioned to convert Greenland from their pagan beliefs. However, the room was 6.5 feet wide and 11.5 feet long, with little room for worshipers.

This author's opinion is that this was originally the "budir" ("boother") of Erick the Red. This word, which we applied to Leif's camp at Vinland, was most probably a cultural icon with the same fabric roof and elaborate carved entry staves that are used on board ship, at home with turf walls, and at the "Things" or regional parliament structures. The size and quality of this roof as well as the general prosperity of the budir in sum represented the stature of the resident in Viking society.

Thjotjilde may have inherited it from her husband and may or may not have used it as a church. Today, the bones of those three interred there are in laboratory shelves in Copenhagen.

During the excavations of this room and all of its surroundings in the 1960s, Danish archaeologists uncovered 144 skeletons. Most of these indicated tall, strong individuals, not very different from modern Scandinavians. One male skeleton was found with a large knife between the ribs, evidence of violence. A mass grave south of the main church contained 13 bodies. According to Neils Lynnerup of the Panum Institute of the University of Copenhagen who performed forensic work on the remains, the bodies were male, ranging from teens to middle age, with head and arm wounds suggesting they may have died in battle.

Early in the twelfth century, the Greenlanders sent a leader, Einar Sokkason, to Norway to convince the king to send them a bishop. Despite their somewhat violent lifestyle, they provided Bishop Arnald with a fine farm on Gardar, not far from Brattahlid. Here they erected a cathedral of red sandstone and dedicated to the patron saint of seafarers, St. Nicholas; with a meeting hall capable of holding several hundred people; a large barn for 100 cows; and tithe barns to contain the goods that would be religiously collected from the farmers by priests and set aside for Rome.

The literature on the subject of the location of Leifsbudir is voluminous, filling whole sections in libraries, the theories abounding, "replica" voyages made, seminars, college courses, theses, theories, articles, books, pamphlets, speeches, talks, arguments, etc. All have filled the void of mystery felt by generations of historians and Scandinavians for whom the old tales of the heroes and the voyages thrill the hearts and resonate in the intellects of men.

So far, none of these studies has been accepted by all persons, or even a substantial portion of them. The site of L'Anse aux Meadows in Newfoundland has come the closest to acceptance by being demonstrably Norse. As time goes on and comparisons are made against the sagas, that site has become embroiled by doubts about whether it was Leifsbudir.

Certainly, that Canadian locale is in a straight line to and from Brattahlid in Greenland. It would entail only a few days' travel thereto in our time, but not the long and elaborate voyages of these epics. Therefore, the search for Vinland has lasted for a very long time.

The Vikings were in America on numerous trips and various lengths so that intermarriage is as likely here as everywhere else they went. Therefore, I wanted to study the Indian culture in the area that I believe was Vinland. Fortunately the famed 1524 explorer Giovanni da Verrazzano contributed to this solution, as did seventeenth century Roger Williams.

London-born Roger Williams was educated in Cambridge, England, a maverick minister who respected native persons and was an early chronicler of the site environs. He was later Governor of the State of Rhode Island and Providence Plantations, an initial champion of religious freedom, and a champion of the Native American tribe.

Many other historians of the area contributed, such as Connecticut colony visionary Governor John Winthrop, a physician and owner of an 8-inch telescope, and contemporary of Roger Williams, and eminent Rhode Island historian Sidney Smith Rider.

I must admit, however, that I have found also a most peculiar response among members of higher education in matters pertaining to pre-Columbian contacts in America. Even though fairly well documented, the suggestions of voyages to America from Europe and Asia before 1492 are relegated to reserved footnotes or omitted altogether. A result of this is that academic "citations" are rare. I invite them to re-examine data along with scientists armed with additional biogenetic capabilities.

What the Vikings Left Behind

How else might we know where Vikings landed? There is a common trait among all peoples to mark their territory and to communicate with each other by markings and other methods. These markings may be cave art, rock art, signatures, cairns, flags, mounds, structures, and even architecture. What can we learn by looking for such Viking markings?

Dighton Rock State Park is in Massachusetts but is only 30 miles from Newport, Rhode Island, near the Narragansett Native American stronghold. Cryptic inscriptions on Dighton Rock have been of interest since around 1680. This is possibly the most studied inscription ever found in North America. Among the usual Indian petroglyphs or rock art on that rock, there seemed to be possible runic symbols. Without any satisfactory conclusions as to their origins, the Rhode Island Historical Society photographed and sent pictures to Professors Finn Magnusson and Carl Rafn to decide whether it was runic writing.

Rafn believed that he could see the name of THORFINN, and the figures CXXXI, which he thought was how many people sailed with Karlseffni. They also sent Rafn descriptions of several other sites similar to Dighton, such as the Portsmouth and Tiverton Rocks to compare. Of course, it was no fairer to ask for decipherment without seeing the rocks than it is to describe a person's symptoms and ask a doctor to diagnose without seeing the patient.

Another scholar, Dr. John Danforth, examined the Dighton Rock and wrote in 1680 that the "uppermost" of the engraving represented rocks and rivers six miles below Taunton. He believed the first figure represented a ship without masts being cast upon the shoals after a wreck. He thought the second figure represented the head of land or "possibly a cape with a peninsula, hence a gulf." It is unknown whether anyone who had studied Indian petroglyphs was asked for their opinion.

An interesting note is that Dighton was the home of famed Indian leader Massasoit, who may have been a descendant of Vikings.

Even Dr. Cotton Mather took a shot at the inscription in 1712. He copied it somewhat incorrectly and sent it to the Royal Society of London, who had nothing to say about it.

Colonel Garrick Mallery produced a two-volume work for the U.S. Government, with many renditions of inscriptions on Dighton Rock in *Picture-Writing of the American Indians*. It was printed in 1893 and was included in the "Tenth Annual Report of the Bureau of Ethnology to the Secretary of the Smithsonian Institution, 1888–89." Mallery was asked to become ethnologist for the military to record Indian information before it was further destroyed. John Wesley Powell was so impressed by Mallery's work that he invited him to serve on the Bureau of Ethnology.

Edmund Burke Delabarre studied the inscriptions from 1918–1927, and believed that he saw the name of the Portuguese explorer, Miguel Corte Real, and the date 1511 (which he believed was when Corte Real came to this area.) However, it was clear from the photographs of that time that people kept adding things to the rock during the early 1900s so that it looked much different than it was originally. Graffiti, alas!

Gaspar and Miguel Corte-Real (sons of Joao Vaz Corte-Real) were the first Europeans to set foot in New England according to some theories. The two brothers sailed from Lisbon to Greenland in 1501. When they couldn't reach Greenland because of ice, they sailed to Labrador. They then went on south to Cape Harrison, Sandwich Bay, and Newfoundland. Miguel sailed back to Lisbon but Gaspar continued his explorations. When Gaspar did not return, Miguel sailed back for him. Both ships became lost.

Some propose that the Dighton Rock in Massachusetts has some communication from Miguel. The petroglyphs and inscriptions were interpreted as a strange Portuguese script that read, "Miguel Corte Real by the will of God here chief of Indians, 1511." Leo Pap in *The Portuguese Americans* proposed the theory that Miguel was shipwrecked in New England and became chief of a Wampanoag tribe and that a crewmember inscribed the rock.

I will finish the story about Miguel Corte-Real. When he returned to Lisbon in 1501, he took 50 American Indians as evidence to King Manuel I that he had been in different lands. Examining the captives, Corte-Real found to his astonishment that two were wearing items from Venice: a broken sword and two silver rings according to Charles Mann in the fascinating book *1491*.

We know about these captives from a letter written from Lisbon by Italian Alberto Cantino. The letter of October 17, 1501, and a map of 1502 were sent to Hercules d'Este, Duke of Ferrara, of the powerful Italian Este family. The map is on display at the Biblioteca Estense in Modena, Italy, according to Manuel L. daSilva, M.D., a Rhode Island internist and archaeologist who wrote "Cantino Spy Letter." Spies were important because various European countries rushed to claim lands and riches before each other. The letter contains one of the earliest descriptions of Native Americans. Here is an excerpt:

> *It is now nine months since this most serene king sent to the northern part two well-armed ships ... On the 11th of the present month one of them returned ... They kidnapped nearly 50 of the men and women of that land by force, and brought them to the king ... they are bigger than our people, with well-formed limbs to correspond. The hair of the men is long ... Their eyes incline to green ...*

After Miguel Corte-Real deposited his captives, he took off again in search of Gaspar, never to return.

Even some Indian petroglyphs (rock art) could possibly have significance. A "shield" motif is featured in some Rhode Island petroglyphs. Even though Vikings always had shields (and shields are much mentioned in the sagas) so did most Amerind groups. This subject is discussed in an 1837 paper by Carl Christian Rafn and in an article, "The Shield Motif in Plains Rock Art" by David Gebhard in 1966. There is also an engraved shield figure found on a stone gorget or hammerstone near Warren, Rhode Island. Variations on the shield figure have been engraved in the so-called Written Rocks at Tiverton, in east Rhode Island, to which an age or origin has not been established.

In addition to these various rocks is a structure that has caused many to look for Viking origins. A stone tower stands guard over the harbor of Newport in Rhode Island. An octagon within a circle, the round stone tower has eight arches at the ground level and two raised stories with curiously placed slit windows. There since colonial times, it might have been a grain mill, a windmill, a Freemason temple, or a remnant of the pre-colonial Norse. Some believe that Giovanni da Verrazzano, who explored Narragansett Bay in 1524, saw it because he used the word "refugio" in describing the site. Furthermore, Gerardus Mercator's map of 1569 shows a tower somewhat near that location.

William Coddington, John and Jeremy Clarke, William Brenton, John Coggeshall (to whom I am distantly related), and Thomas Hazard founded Newport on April 28, 1639 and set up a plantation at the southwestern end of the island called Aquidneck. Those early settlers left no records about a tower. The first person to mention it was Benedict Arnold (great-grandfather of the infamous turncoat Benedict Arnold).

Arnold was born on December 21, 1615, in Ilchester, Somersetshire, England, and moved to Rhode Island. On November 19, 1651, he settled in Newport. On May 19, 1657, he succeeded Roger Williams as President of the Colony until the arrival of the New Charter from King Charles II in 1663. He was named the first Governor under the New Charter. The hill where the tower stands was purchased by Benedict Arnold in 1663 and became known as Arnold's Farm. He died on June 19, 1678, in Newport and is buried in a plot "lieing between my dwelling house and my stone-built wind mill" as he requested in his will dated December 24, 1677. The family cemetery was at the southeast corner of the farm, at what is now 74 Pelham Street in Newport. In his will, Arnold appeared to be taking credit for the "stone mill" but many have wondered whether he built it or just bought land where it was already present.

One confusing feature is the story that the tower contains a stone with a cryptic engraving. The engraving appears to have five figures, which some call Norse letters HNKRS. Some say those letters could represent the Old Norse word for stool, meaning the seat of a bishop's church. Others say that it could stand for the name of a bishop who is recorded to have sailed to Vinland one or more times. His name is Bishop Erik (Eirik) Gnupsson who was known as Henricus, hence HNKRS.

An excavation at the end of 2006 concluded that the Newport Tower was, in fact, no older than the time of Benedict Arnold and was probably built by stonemasons he employed. I will add something more about this excavation later.

Let us consider the Kensington rune stone site in Minnesota. That is not so very far away from Viking explorations if they sailed against the St. Lawrence River current and portaged around falls to get into the Great Lakes area. In 1898, Olof or Olaf Ohman, a Swedish immigrant working his fields in Douglas County, Minnesota, found a 200-pound engraved stone. Douglas County has the eastern Chippewa and Pomme de Terre Rivers and is not far from the westerly Mississippi River. He said that it was tangled in the trunk of an uprooted

Aspen tree but he thought the engravings were somewhat familiar. Controversy has raged over the Kensington rune stone's authenticity. It is reported to say:

> Eight Goths and 22 Norwegians on a journey of exploration from Vinland very far west. We had camped by two rocky islands one day's journey north from this stone. We were out fishing one day. After we came home, we found ten men red with blood and dead. AVM [Ave Maria] save from evil. Have ten men by the sea to look after our ships fourteen days' journey from this island. Year 1362.

Many considered this a hoax created by Ohman, a proud Swede who might have wanted to prove that his people were there earlier. Scott Wolter and Richard Nielsen have now presented a new approach to this and additional research in their 2005 book, *The Kensington Rune Stone: Compelling New Evidence*. They point out that Ohman used the stone as a doorstop and only his son thought to describe it to others. They add that a supposedly made-up letter (j) in the runes, which caused linguists to scoff, has now been found at least eight times in the *Codex Runicus* from around the 14th century. They point out that although someone may have re-encised the runes in recent times to make them more readable, other markings and glaciations suggest that the stone was underground for a very long time.

Questions remain about what the runes described. One possibility comes because King Magnus, King of Norway and Sweden who ruled the latter country from 1319–1363 ordered Paul Knudsson or Paal Knutssen, a judge and member of the Royal Council, in 1355 to take people to Vinland to restore or insure the continuity of Christianity to the settlers. Weather delayed him but he sailed to Vinland in 1356 and returned by 1363.

Ivar Bardson, a priest from Greenland, had sailed with him as well and finally arrived back in Gardar. Friar Nicholas of Lynn, a friar who possessed an astrolabe, reported on the journey in 1363/4 in the *Inventio Fortunata* and described what sounded like a venture into Hudson Bay from which not all returned.

R. A. Skelton *et al.* described the *Inventio Fortunata* in their 1965 book, *The Vinland Map and the Tartar*. Some, including Wolter and Nielsen, think that the Kensington Rune Stone could be a faithful rune describing problems of this group in 1362. However, its location is so far from a river that the message does not seem to fit.

The great volcano Oreaefajokul erupted in Iceland in that year, 1362, and obscured the area with pumice, darkness, and pollution for months. It is possibly worth noting that not only had the volcano devastated Iceland and wiped out some settlements in 1362 but also the Black Death came to Greenland only seven years later in 1369. Many might have fled Iceland and Greenland at that time. Some may have returned to Norway, Sweden, Denmark, or Europe, but others may have sailed to America.

Carl C. Rafn listed eight runestones along the coast of southern New England that had been described to him by New England historians well aware of the legends of Vinland. Recent research has discovered a few more. Most have been lost, destroyed, or defaced with graffiti, but two are still extant and one is preserved in a museum in Dighton, Massachusetts (a place with Narragansett Bay access.)

Another apparently known to Rafn was on the isolated island Nomans Land Island, near the western shores of Martha's Vineyard Island. The public is not permitted to use that island because of unexploded shells left over from World War II when it was used as a gunnery range.

The owner of the island, Joshua Crane of Boston, discovered it in autumn of 1926. It was soon in the newspapers and an article was published in the *New England Quarterly* in 1935 called "The Runic Rock on NoMan's Island, Massachusetts." The article by Edmund Delabarre (who went to examine the rock) began:

> The discovery a few years ago, on a bleak part of the shore of Massachusetts, of a rock bearing the name "Leif Eriksson" in unmistakable runic letters and the date "MI" in Roman numerals ... has been described fully by two writers, Mrs. Wood and Mr. Gray.

He was referring to a book written by Annie M. Wood in 1931 called *Noman's Land, Isle of Romance* with a chapter on the rune stone by her husband, Captain Cameron E. Wood and *Leif Eriksson, Discoverer of America, A.D. 1003*, published in 1930 by Edward F. Gray of the British consular service posted in Boston. This was followed by the military consignment of the island and the opinion that Roman numerals and runes do not appear together so it must be a fake. Little has been done about the rune until 2003 except for finding that Roman numerals and runes do appear together.

Scott Wolter and Richard Nielsen visited and photographed the rock on August 27, 2003. The New England Antiquities Research Association (NEARA) requested them to report on their trip and their findings. Constantly washed by waters and encrusted by sea life, it has deteriorated but they made their report. Their observations and hypotheses are included in their 2005 book, *The Kensington Rune Stone*. Essentially, they say that it could or could not be authentic.

We could describe "Thorvald's Rock" in Hampton, New Hampshire. That old carved rock is now in an enclosure, making it impossible to see the lines supposedly carved upon it. The site does not match the saga's description of Thorvald's last resting place but those who named it might not have an intimate familiarity with the pertinent sagas. We could add another finding of the runic carvings on rocks in Boar's Head, New Hampshire, but they are also indecipherable and match nothing in the sagas.

A few rocks might have some significance. One is called Symbol Rock in Rockville, Rhode Island, which some have attributed to Vikings or Viking influence. It is located in the Yawgoog Scout Reservation. It has petroglyphs (perhaps done by an earlier generation of Narragansett Indians), circles, and an outline of something like a pregnant woman.

Well, all these different versions of inscriptions remind me of Garrison Keillor and his wonderful story about a Viking rune found in his hometown that was translated. It said: "8 of {us} stopped & stayed awhile to visit & have {coffee} & a short nap. Sorry {you} weren't here. Well, that's about {it} for now."

Let us try to portray Vikings as accurately as possible. Neither Northeast Indians nor Vikings seem accurately portrayed in the public mind and it was with some surprise that I noted so many similarities between the two cultures. Indians come through as primitive in the popular view, with Vikings violent and barbaric. Vikings have been portrayed as not only primitive but dumb in cartoons such as *Hagar the Horrible*. They were portrayed as saying things like "Remember: pillage first—then burn!"

Vikings were neither more barbaric nor dumber than other contemporaries were. However, it is true that Erickson and others moved from Iceland to Greenland because violent behavior had caused them (at least Erick) to be exiled. If the United States permitted manslaughter offenders simply to be exiled, we would be perceived as very humanitarian.

Thorvald Erickson massacred two Native American groups caught asleep in Vinland and that would strike us as typical of Viking barbarity. Of course, English colonials did the same thing to many Indians. Yet, there is much more to it.

Thorvald's explanation that the first party slain seemed to them to have been "outlaws from the land" gives us a clue to the more typical live and let live attitude in the Northlands. He was acting according to his concept of right to persecute "outlaws" wherever found. For in Viking society, outlawry was the worst punishment that the community might mete out, but even so, that was often tantamount to death. Anyone—even apparently of the slave class—had the right to slay an outlaw if they chose. The outlaw had the option of a "running start" and if he had friends or means of escape, might survive and re-enter society later.

Speaking of the slave class, the Vikings in the sagas were most lenient in their treatment of slaves. It was a common thing for the Vikings from Sweden and Denmark to raid and capture slaves mainly from Slavic-speaking countries according to *The Viking World* by Philippa Wingate, *et al.* Norwegian Vikings took slaves from England, Ireland, the Orkneys, the Faroes, etc. Of all products, slaves were the most valuable commodity. Men were used for labor and women were used for both labor and sex. This was as logical as the use of Negro troops in the Union Army in the American Civil War. A man enslaved might be the better warrior by being closely bound to his owner.

The Norse Vikings took slaves on their small ventures to America. Slaves were more valuable for their particular skills and subjugation was not as prominent a feature as with other Vikings. Slaves on trips to America were not, for example, kept in the galleys or mistreated as badly as others. In Iceland and Greenland, they were almost like Southern household blacks and treated as part of the family. Recall the male and female slaves who were runners and returned after three days with grapes and wheat. If they had been treated badly, here was their chance to escape.

Sometimes Vikings resolved transgressions that were more serious by dueling, but usually by votes and rulings at the "Thing" or courts. If all else failed and a suit was not resolvable, it might default to a "blood feud." However, the purpose of the Thing (parliament) was to try to prevent this. There was also, of course, punishment meted out by chieftains to subjects. Their social life was stable and self-regulating. They were, apparently, within their culture, "law-abiding." When one considers that murder and manslaughter were not punished by imprisonment or death but by banishment, this strikes modern man as an altogether humane form of retribution.

Erick the Red's experience was typical of a ranking chieftain who owned a ship and took the option of exile to explore Greenland. Perhaps his wife's eviction of him from their conjugal bed was the most extreme of religious warfare in the North. There do not seem to be any records or recollections of religious persecution taking place in Scandinavia although that was the norm in the southern climes of Europe.

We can learn something from the preserved bog burial of the Tollund man (whose face looks very much like Swedish actor Max von Sydow) in Denmark. His is possibly the most photographed and studied part of any bog body. He died or was executed about 210 B.C. The archaeologist who studied the body and excavated it from the Tollund fen was Peter Glob, summoned by police who first thought there had been a murder.

Glob wrote a fascinating account of his 1950 find in *Eyewitness to Discovery*.

> *On his head, he wore a pointed skin cap fastened securely under the chin by a hide thong. Round his waist, there was a smooth hide belt ... His hair was cropped so short as to be almost entirely hidden by his cap. The air of gentle tranquility about the man was shattered when a small lump of peat was removed from beside his head. This disclosed a rope, made of two leather thongs twisted together, which encircled the neck in a noose drawn tight into the throat and then coiled like a snake over the shoulder and down across the back ...*

Glob concluded with the contents of the stomach—a gruel prepared from barley, linseed, "gold-of-pleasure," and knotweed. A gruel made to this recipe was served up on an English television program in the summer of 1954 to two well-known archaeologists—Sir Mortimer Wheeler and Dr. Glyn Daniel. Sir Mortimer finished up by saying that it would have been punishment enough for the Tollund man to have been compelled to eat this gruel for the rest of his life, however terrible his crime might have been.

This leather cap, firmly fastened, was a common feature of European medieval dress, and was in many paintings by Peter Bruegel and even paintings of Popes. It served as a helmet liner and was worn by Popes as symbolic of their brotherhood with the common people. The helmet itself was iron, not the kind of material one wants in contact with the brow in freezing weather. The militia, guard, or night watch might be called out at any time, day or night, rain or shine, cold or hot. Perhaps Tollund man was strangled because he failed to respond to a call to arms, or, perhaps, fell asleep on sentry duty.

The Inquisition, persecution and massacre of Huguenots, periodic pogroms of Jews, drawing and quartering, beheadings, and burning at the stake of heretics were typical of more southerly European society all through the Middle Ages—even up to and beyond 1800.

Community-based barbarism of this type did not seem to occur in Scandinavia and it should be noted that the formation of the Democratic Republic of Iceland and its acceptance of Christianity was done by vote at the Icelandic Thing (Parliament) at about the same time as the Vinland voyages. A class of so-called "Barons" had forced the Magna Carta of England upon the English King. In Iceland, there seems to have been a placid acceptance of a status quo by the Chieftain class who actually voted it in. Indeed, those Vinland voyages were essentially Christian endeavors.

Part II
▼
The Vinland Sagas

*The Vinland Sagas entire and as far as we know, exclusive.
"You can't enjoy the game without a program."
This narrative is the basis of our program.*

Introduction to the Sagas

The following documentary is a free translation of the sagas of the north. These stories are not chronicled completely in any one place but are broadcast throughout several "Boks." Of these, there are numerous variations resulting from awkwardness of translation through the various languages or "Runic," to Latin and finally to English, mainly, and other languages only to a limited degree.

There are two prime sources of the epics; several stem from a Greenland base, which survived in the Icelandic dynamics themselves, and were further complicated by incursions by persons of Danish residence during the sovereignty of Iceland by that nation. Many of the documents themselves have been removed to Copenhagen where, for reasons best known to the Danes, they were kept in confidence. All surviving have been returned to Iceland where they are preserved in a museum. Some may have been lost during World War II; one, unrelated and in olden times wrought on sheepskin, was actually cut up and made into a vest for someone who wore it thus, complete with sagas all over—one is forced to read past the buttonholes.

However, the stories have been long transmitted and are well known in Iceland—enumerated, catalogued, and cross-referenced. The major documents are called "Flateyarbok" and "Hauksbok" but there is much more than in these sources; some references are extant in sagas of Norway and Denmark. Though little known, there is even a remarkable quantity of literature in medieval writings and maps, in Bremen, Germany, and in archives of the Roman Catholic Church.

The author makes no pretense to being an expert on the detail and accuracy of the "Runes." Read specifically, their meanings are difficult to extract. Indeed, it took me a year and a half of reading before I was able to 'connect' to the real message, and even at that, someone before me had done the major work. Eventually, I simply read with the object of extracting whatever truth might be found therein. I tried as best I could to 'comprehend' them—to 'milk' them as dry of information as possible.

They are much more voluminous than might be expected. The late Frederick J. Pohl, an outstanding Vinland Scholar, in correspondence to me said that he had identified some 83 or more statements of course changes by the varied ships that had occurred between Greenland and Vinland. I have been unable to penetrate the sagas to that degree but I believe it.

One, which interested me, is that Leif's approach to the island I refer to as "Island of Sweet Tasting Water" was from the northeast. Pohl's analysis was from an objective observation—in some way by either direct statement or implication, he discovered that information. My analysis of it is subjective. I arrived at the same conclusion for this course from analysis of plausible courses to and from the site proposed. The sagas statement of what the course was from that island to the Leifsbudir is clearly stated and is one of the more critical issues of understanding Vinland.

In most cases, the sagas are summarized much too briefly, often to a few paragraphs. They are actually much longer, possibly book length in total, and if footnoted and indexed, they might take up several thick volumes. If this is done, the matter becomes so complicated that the story line itself is lost.

Hence, what follows is a rather straightforward narrative of the Vinland sagas made as readable as possible, and to the length necessary to the problem at hand; namely, discussion of a specific site as probably being Leifsbudir.

Some of the factors addressed in footnotes are direct extractions from the sagas, others from prior students' deductions, some from alternative sources, and one, my own humble contribution. For instance, the presence of maple trees in Vinland is not described directly in the sagas but is from a later tale of Thorfinn Karlseffni while he is in Europe and mentions "mossur" (Old Norse Maple) in a business deal there. What you will read here is controversial. No two readers seem to arrive with the same understanding but this is the more commonly accepted version of several possible sagas with one exception: a matter debated in several forms, of which this narrative selects a neglected, often over-looked, but acceptable version. Of the alternative versions at that point, this one has the virtue of ultimately bringing the ship of our explorations into a harbor that now seems known to us—Leifsbudir.

The Saga of Bjarne Herjolfson

The first tale can be touched on lightly. It is one of a ship captain named Bjarne Herjolfson and both Bjarne and his father, Herjolf, figure in it. Herjolf is with us to this day, for the southernmost cape of Greenland is now named "Cape Herjolfness." There is said to be much detail in this saga because the adventure is frequently cited as almost "log-like," with courses and sail changes documented. In brief, however, the matter begins with the thought in Bjarne's mind that he wishes to spend the Yule with his father who has moved to Iceland some time previously. Where Bjarne was when this filial thought struck him, we do not know, but we presume it was in Norway. It was 986 AD and we can guess to have been rather late in the year, if Christmas was in the offing. He sets out and makes Iceland with no particular difficulty, but there he finds that Herjolf has moved on to the new, outlying colony of Greenland, founded by Erick the Red some six years previously.

Evidently eager to join his father, he polls his crew remarking on the dangers of sailing unknown seas without a pilot and the unified band sets out for the distant land. In some days, they come within sight of it—or a land they mistake for it—when they are struck by a severe storm and driven before it for many days. They are storm and fog-bound for "a long time"—some say for fifteen days—before the weather clears and they again establish their bearings.

When they do, they discover that they have been driven very far south and so set course due north—apparently by the reasoning of ancient navigators to "run the latitudes," not having any accurate means of determining longitude. In one day's sail they make a landfall and they describe it as low and wooded with many dunes. The crew asks Bjarne for permission to go ashore but this is refused.

They make off again and in two days' time make another land which is also described, this time as beached, hilly and forested. Again, the crew entreats to go ashore and again they are refused.

Sailing northward yet, in three days' time they come to a third land and this is said to be bleak and barren with a wide plain and mountains with seeming glaciers off in the distance. The suffering crew again asks to go ashore, perhaps with some insistence, because Bjarne replies to them with some asperity that it is refused and even adds to the refusal the astonishing remark that they "... still have sufficient in supplies." That is a good indicator of Norse seafaring and endurance.

Sailing again, they are away from land four days and this time make a landfall that they identify as, and find to be, Greenland. They make into a fjord there and see a small boat pulled up on the shore with a man standing beside it. Closing the distance, they find that the man is none other than Bjarne's father Herjolf. It was said that their Yule together was among the happiest ever among anyone.

This is an important saga for a number of reasons. First is that Bjarne's action in polling his crew before starting the voyage to Greenland is most significant in gaining insight to the times and the seamanship. His remark of refusal in saying there was still sufficient in supplies adds to the implied high level of discipline that must have been in operation on board. The crew may have numbered anywhere from 8 to 15 persons.

But most important is the later information that this was the very ship that Leif Erickson bought from this very ship captain fourteen years later for his own epic voyage. With the ship, we can suppose he also obtained the crew, some of whom were the same as before. When Leif did set sail he did so with the stated purpose, several times reiterated, that he intended to reverse the courses of Bjarne. So far as is known and demonstrated, he was successful in doing so. Hence, the description of the land Bjarne saw first is the same as the one Leif found last and which became the sub-coast of Vinland, "... low and wooded with many dunes." It entailed three sea passages, one of four days, one of three days and one of two days. Many later recordings of this say that "... it was not far from Markland to Vinland"—a two day passage.

The Saga of Leifur Eiricksson

The early stages of Leif Erickson's saga give insight to the man who commenced the proceedings. For an individual who lived most of his life in a "lost colony" a surprising amount of information comes down to us through history; even his very words, his personality, his coloring, his "modus operandi." He was much admired and a hero in his own time. One feels somewhat more than the dutiful obeisance of the numerous serfs, vassals, and slaves that made up the bulk of the population of the day.

He was said to have been tall and powerful but fair in all his dealings with others. A skeleton has been exhumed in Greenland, which has been thought to be his. The possessor standing in life was about 6 feet 2 inches tall—immense for the time when the average Scandinavian, while still larger than his Southland kin, was considerably smaller than today.

He was master of the home "Vik of Brattalid" in Greenland for many years after his voyage to Vinland. It is not known whether he ever returned to his epic discovery even briefly, and yet he is seen to be most seriously possessive of the place. He was several times importuned to sell it but every time refused to do so although freely loaning use of it to others. He lived at least to his forty-fifth year.

The first we hear of him was when, in his early twenties, he made a voyage to Norway, apparently for the purpose of being presented to King Olaf Tryggvasson who had but recently forged one of the first of sizeable Norwegian Kingdoms. Leif was in for a rude surprise, for King Olaf was an ardent convert to Christianity and Leif was still a follower of his father Erick's pagan beliefs. King Olaf had a set policy at this time in that all of his followers must convert to Christianity or be executed.

This rather shocking directive in behalf of the "Prince of Peace" was given to Leif as well; he was granted a fortnight to make a decision. From what the sagas say, Leif considered himself a noble warrior. He demonstrated his Norwegian heritage by this attitude, for you cannot 'crowd' a man of Scandia, took the entire allotted time to make up his mind. He eventually accepted baptism—we know not with what inner feelings—into the Christian faith and was given the assignment to return to Greenland and convert the entire colony there.

Earlier, on his eventful trip towards Norway, his ship had landed for an apparently extended waystop at the Faroe (var. Orkney, Skilly) Islands. There, he meets and becomes enamored of one fair Thorgunna and at his departure, she pleads with Leif to be taken along. Leif refuses. An interchange between them is retained in the sagas as a matter of considerable interest and reminder that human nature, after all, has not changed very much over the centuries.

Leif:	"This cannot be (to allow Thorgunna to accompany him). I have not the wherewithal to deal with your relatives, and we so few in number."
Thorgunna:	"I wish you to take me in any case, with their permission or without, for it is my desire to make my life with you."
Leif:	"I cannot take this risk and must leave you here."
Thorgunna:	"You will find that this is not a wise decision and it will bring regret to you."
Leif:	"I will put it to the test notwithstanding for I must leave you here."
Thorgunna:	"Then I must tell you of the importance of the matter for I am with child and you are responsible. Leave me if you must but I tell you I will follow you to Greenland."

The interchange ceases here but must eventually have been amicably concluded for it is said that Leif gave Thorgunna a gold ring, a belt made of walrus tusks, and a "wadmal mantle" (evidently a cape made of some specific coarse cloth).

While this tale takes us even further from our goal, the sagas intimate that Thorgunna did, indeed, reach Greenland and that the issue of the two was a boy who was named Thorgils. It was said of him that there was something about him that was "—not quite right"—in any event he died young.

Leif's trip back to Greenland is not chronicled but events at Brattahlid that unfolded give considerable insight as to what happened next. Recall that Leif had been given the mission to convert the settlement to Christ and, evidently, he applied himself conscientiously to this task. He is immediately successful in converting his mother Thjothild and certain others of the Vik settlement, but his violent and combative father Erick resists mightily. Thjothild enters the controversy and resorts to the age-old tactic of evicting Erick from the conjugal bed with the admonition that he will not be welcome back until he decides to convert. Far from doing so, he carries off his own cot and takes to a hut at the far end of the settlement, doubtless to brood on the injustices wrought by fate and prodigal sons.

Erick is not the only one who resists with vehemence. Others—presumably most of the males—do also. Viking warriors take great comfort in the pagan warlike gods so encouraging to loot, murder, and rapine. In brief, the settlement of Brattahlid is riven with dissension in short order. Leif lost much in popularity; in fact, it is implied that he received more than several death threats!

His solution to the problem is one which has occurred to many in duress, but Leif had the opportunity to exercise options in the purchase of the ship of Bjarne Herjolfson for a good and true Viking voyage. The dissension with his father is evident in the use of another ship than Erick's, but on the other hand, it is also said that Erick was originally scheduled to make the trip as leader—either figurative or actual. Some of Leif's arguments to Erick give insight to the occasion for Leif flatters his patriarch in several ways, saying *"... there are none so good at handling the tiller (steering handle)," "... you are seen to be lucky, and more men will follow you-,"* etc.

Evidently, the importunities of Leif were initially successful in overcoming Erick's reluctance, but on the day of departure, Erick, in riding to the moorings, was thrown or fell from his horse. He takes this as an omen, declaring that it signifies his explorations are over, and he decides to stay home.

Departure

Leif has successfully negotiated for purchase of the ship of Bjarne Herjolfson. They organize, make ready, and start the epic journey with a crew clearly specified as thirty-five men. We can assume that this start was in accord with Viking custom, which was to commence as early in the spring as possible. In those days of somewhat easier climate, it could have been late in the month of April or early March, I suppose.

After some time at sea, they espy a rugged, mountainous coast and make for it, landing for explorations. After a few days, Leif gives a speech and says, *"Methinks this is the land that Bjarne saw at the last of his journey. But whereas they did not land, we have done so and explored, so the claim for it is ours. Still, now that we have seen it, I can see that it seems to be good for nothing (the land) and it is my wish to go on to the other lands."*

They have caught salmon in the streams and they describe the land as having flat stones. The stones are variously described as being a single, large, flat stone; a number of large flat stones; the entire plain between the shore and the distant mountains being strewn with large flat stones; flat stones about 12 "els" across (18 ft., or 5.5m); or flat stones of a size that two (var. three) men head to foot could not span them. Leif names the place "Helluland," meaning "large flat stone (or 'slab') land" and they depart, still on the reciprocal of Bjarne's homeward trip. The sagas mention westward-tending courses from this point.

They come on land again and this time on terrain described as hilly and well forested and so they name it "Markland," meaning "forest land." Leif again is quoted as saying that this seems to be the second land that Bjarne saw and they again push on in their journey.

Eventually they see the third land and Leif declares it seemingly to be Bjarne's first landfall after his storm-tossed epic. Along the route, they have passed what they name "Wunderstrands" (also "Furdirstrands"—meaning far along beaches) because, as it is said, it took so long to sail past them.

The atmosphere of the saga is dreary—beaches signify danger, the dreaded "lee shore," to seamen. In later coasting, they describe the land as being forested and the shoreline well broken with many bays, rivers, and estuaries.

They come then to an Island and land on it. They said there is "land" or a "mainland" to the south of it and that there is a spring or source of water *"... down there in the grass";* they *"... dipped their fingers in it and tasted it and thought that they had never tasted anything so sweet."*

They leave this island and most fortunately for us define their course precisely as being *"... north to a fjord across that sound which lay between the island and that cape which pointed northward,"* also translated simply as *"... more back toward the north-pointing cape."*

A river is reached and a landing attempted. Rare sources specify a turn *"west around a headland to where the river flowed* (var. fell, tumbled) *out of a lake into the sea."* The ship grounds out on landing (becomes immobile) and the men rush over the side and wade ashore because, it is said, "*... they could not wait to explore.*" This river mouth is described partly in situ and partly through references in later events:

It has a sand or "gravel" bar across the mouth, it has white sand beaches, it is impossible of entry except at high tide. The forest came very close to the sea. When the tide went out "*... it is a long way from the ship to the sea* (meaning a long tidal "fetch").

Additionally, we can assume that it was an attractive place to them if they rushed ashore so enthusiastically; and a "safe" bottom as well—doubtless the sand bar. While they are all off exploring, the ship refloats unattended and some of the crew must row to it in the afterboat to recover it.

They then rowed the ship up the river into a lake and there they moor for more explorations. After a very brief stay, Leif declared that he wished to build "houses" with intent to form a permanent camp and these are—or at least his is—built on heights overlooking the lake.

He then gives a most significant and informative order to his crew: *"We will divide into two watches; one will stay with the ship and the other to explore. But each party must remain together at all times, neither must they go so far that they cannot return to the ship the same day. The watches will alternate and I (Leif) will accompany either party as I so wish. Those at the ship will build houses and cut timber and vines* (var. 'for cargo')."

Possession

At this stage of the sagas, several references are made concerning the whereabouts and climatic conditions of the place. These seem to be an attempt to define the place as clearly as possible to establish Leif's claims which, though now scattered through several sagas, are complete enough to establish certain rights to discovery within his culture.

He has defined his courses (south with westward tendencies); he has landed, named, and explored three major landfalls en route (Helluland, Markland, Vinland); described pertinent features of Vinland (low and wooded, broken shore with many bays, rivers, estuaries, and one fjord) and the all-important offshore island of military considerations; precisely defined the course from this island to his major landing; and described the site itself. Other items are sometimes intermixed with observations from later travelers so I can best leave these to a separate section of clues relating specifically to the coastline and Leifsburdir.

Two major items, however, belong subsequently to the following episode, which seems to be the only adventure taking place as they settled in for the remainder of the season. The duration of the trip, itself, is not mentioned but must be assumed to have been distant if the ship stayed the full year. Common sense dictates that the prudent explorer would set appropriate plans or bend his courses homeward by about the time of the summer equinox, certainly not much later.

One day they went out from the house and Leif accompanied the party which went exploring. But as they were returning (var. after they had returned; as they were about to return), they miss one of the men. This was Turkur, a small and insignificant looking man who was highly skilled at crafts and military arts. He was *"... ill-visaged, with bulging brows and protruding eyes but skilled in crafts and martial arts."* They immediately set about as a search party but they have not traveled far when Turkur, who had gone on ahead, appears to them acting strangely and mumbling in his own tongue. Leif addresses him severely, *"Why have you separated yourself from us and remained away so long against my orders?"*

For a time Turker cannot reply. He makes grimaces and talks rapidly in his southern tongue, shrugging his shoulders and blinking his eyes. After a time he seems to recover and says, *"I have not gone much further than you but I have found that which you have not seen and might find difficult to believe. I have found grapes and grapevines!"*

Leif replies, *"Can this be true, foster father? Are you sure they can be grapes?"*

Turkur responds, *"Yes, it is certainly true; they are grapes and I know them for there is no lack of grapes in the land where I was born!"*

The sagas then go on to say, rather cryptically, *"... and they filled the afterboat with them (the grapes)."* These few words, seldom addressed, are remarkable in themselves if you know something about grapes, wine, and small boats.

(In my final analysis of the site they stormed into my consciousness for the conclusion that I must proceed with the study as best I could. The mystery of this unusual statement was resolved at the site.)

Two of the clues of location of Leifsburdir seem to have been made by Leif and these two are a remarkable solar observation and a mention concerning the climate.

It is said of Leifsburdir that *"... the days were more equally divided than in Greenland"* so that, *"... on the shortest day of the year the sun was already up at 'Dagmalastad' and still up at 'Eyktyarstad'."* These two times of day seem to correspond roughly with breakfast time and dinnertime and are not now precisely known, but many analysts opine that they were 8:00AM precisely, and between 3:00 and 4:00PM (less exact). This factor has been finely analyzed by others and will be addressed later; however it is read, the factor does signify great range from Greenland.

The climate is said to be such that *"... the grass hardly withered; there was no frost so that it seemed the cattle could winter outdoors without shelter* (var. fodder)." The epic then goes on briefly to say that the next spring, *"... they made ready and sailed away"* (to Greenland).

But there is one episode described on the journey home which happens to yield many insights as to Leif's luck, his personality, his age, and certain factors of timing and distance to Vinland, even though, almost certainly, it occurred on or near the coast of Greenland itself.

As they are sailing along, Leif himself is steering and someone among the crew questions his courses, querying, *"Why do you steer so far to the north?"* (var, '*off course*').

Leif replies, (one can sense a certain testiness) *"I have a mind to steer my ship as I see fit but I do think I see something to steer for. I think I see a ship."* So, everyone looks out and all they can see in the distance is a skerry (reef). But they continue towards it and eventually they make out that wrecked upon the skerry is another ship, which Leif has seen from much further off than anyone else. Moreover, they find that there are survivors and these are taken aboard and rescued. Leif agrees to take them to the settlement but claims salvage rights to the wreck, as has been the custom of the sea from time immemorial. The ship had been laden with timber, among other things, but Leif already has a rich cargo of timber and "vinber" (and wine?) from Vinland so he must leave the wreck for a later time.

At the return to Brattahlid, they are hailed and welcomed. Leif is given the sobriquet of Leif 'the Lucky,' because he has returned safely and with a huge new land to claim, two fine cargoes, and, from what we can detect, resolution of his personal problems. It may be that he married and followed that peculiar custom of Vikings of often receiving inheritance at marriage instead of the patriarch's death. At any rate, he seems to have settled down in Brattahlid in Greenland rather than his other home in Vinland. He became the leader or 'Jarl' (Eng. 'Earl' of the settlement of Brattahlid at death of Erick, said to have occurred at this same year of return.

As far as anyone knows, Leif never returned to his far-off discovery. One hopes that he spent some time and offered some security to the unfortunate Thorgils, his son. He appears again only once in the sagas, much later, in yet another insightful episode described which will be told in due time.

The Voyage of Thorvald Eiricksson

The tale goes on with the story that Leif had a younger brother named Thorvald. Thorvald seems to have been obsessed with that tension so common in younger brothers of fortunate older siblings. During the following winter when all are congregating in the long hall for sociability and tale telling, Thorvald takes Leif to severe task for not exploring Vinland sufficiently. Eventually, this situation is resolved after some negotiations with the arrangement that Thorvald would return to Leifsbudir and explore to his heart's content. He will be free to borrow but not to own Leif's houses but, moreover, as part of the agreement, Thorvald will salvage the cargo of the wreck on the skerry before he departs on the longer journey. This he does and then leaves with an apparently indeterminate size of complement. The figure may have been as low as 30 men, or as high as 65.

(This transcript was written very early in my research. Subsequent readings of *Flaytybok* in direct translation finds that the crew was stipulated precisely at 30 men. In fact, it is now apparent that the saga complements of all the expeditions are exactly defined.)

He uses the same ship that Leif sailed, which now, as we see, has been to the coast of Vinland three times when, at last, Thorvald arrives there. There is little said concerning his outward journey; only that he arrives at Leifsbudir without difficulty, occupies his brother's houses, and commences to explore. His stay and the explorations are indistinct, with an aura of mysticism and indefiniteness that make his entire stay at Leifsbudir almost dreamlike. The only episode that comes through with any clarity is a journey to the west of Leifsbudir, stipulated as having used the afterboat rather than the ship. He finds the coastline wooded to a rocky shoreline, eventually reaches shoals with difficult navigation and only one chronicled sign of men, this being a structure of some sort on an island with no one about. The structure is variously translated to be a barn, shed, granary, or corncrib. After this, they return to Leifsbudir. They are on the coast for two years (actually, by analysis, a year and a half, for we must account for a period of delay for the trip down), and then commence their homeward journey which in detail and drama add much to our suppositions.

As they start homeward, they make a landing on a strand as they see something on the shore that interests them. It is three native boats upturned and beneath are five natives asleep. Viking blood comes to the fore and the sagas simply say that these natives were killed. Uncharacteristically for Vikings, who were not in the habit of explaining themselves, they rationalize that the reason they did so was because these men seemed to *"be outlaws from the land."* Thorvald, too, may have been converted by force, as it would not seem his Christian ethics ran very deep! The native food supply is described as being *"blood and marrow mixed and carried in sacks."*

They go on and make another landing on a cape or peninsula which is so covered with the excrement of "animals" that it is hard to walk there, so they name this place "Dungeness"—cape or headland of excrement.

Again, while sailing on, perhaps in a storm, they suffer a grounding and this is severe; it breaks the keel of the ship in such a way that they cannot proceed without repairing it, which they do; but these repairs consume two whole months. When it is finished they set the old keel upright in the sand, some say in the form of a cross. This would be a natural thing to do as the broken keel was in at least two parts. They name the place "Keelness" cape or peninsula of the cross.

But even though these repairs must have been aggravating, we can see from immediate events that Thorvald was yet in no great hurry to sail directly to Greenland for his courses from Keelness are well defined and signify an attitude of continued explorations. He is said to have sailed north and around a north-pointing cape and thence turned inland (var. west, occasionally east, "inland" the more common translation) and then comes to a place which is well detailed and where an informative drama takes place.

They enter a bay (var. fjord) and make landing on an island or headland. It is high and has steep banks so rather than make the ordinary "beaching" landing, they moor alongside the bank and go ashore on a gangplank, then climb to the top of the hill. Off in the distance they see "mounds" which seem to be a village of natives. Then they see closer to, and on the same island, more upturned native boats so they investigate and find yet nine more men asleep under them. Again they attack the innocent natives and manage to kill eight, but one survives to escape by furiously paddling a "boat" away toward the village.

They return to the top of the hill and Thorvald looks around and declares that the land seems to be so fair and he likes it so well that he would like to stay and settle the place (var. *"... linger there a long time"*). Thorvald and others of his crew then make speculations on where they are. They say that it seems they are not far from their old camp, Leifsbudir (var. Hop), and that the mountains seem to be of the same range as those at Leifsbudir. (This remark is the most important in the sagas, for it enables comprehension of the land of Vinland.)

Suddenly, everyone is overcome with a mysterious lethargy and they return to the ship (or a nearby meadow) and go to sleep. They are shocked awake by a voice that comes from above, saying, *"Arise, Thorvald, you and all your men! Danger is at hand! If you would save life, take to your arms!"*

Indeed danger was close by! They look out and see a multitude of boats approaching—it is the survivor of their massacre returning with many reinforcements. The "skraelings" (skreetchers, barbarians) do not stop to parlay but immediately go into attack while still in their boats. In the interval while the Norse prepare to defend themselves, Thorvald gives orders that they should fight only so hard as to defend themselves and not go into counterattack. For, he says, *"I still think this so fair a land that I would settle here and I wish to make no more enemies for this combat."* (Translation approximate.)

The battle is fast and furious and a number of natives are slain. They eventually break off and return the way they came, back toward the distant village.

Then Thorvald turns to his crew and inquires if anyone is wounded, and is told that no one is hurt. He then says, *"I fear it is as I said and that I will linger here for a long time, for I have been hit with an arrow and I believe it will be the death of me."* He then pulls the arrow out from under his armpit (var. groin) and shows it around. He then says, *"This is indeed, a fair land for I see much fat from my insides stuck to the arrowhead. When I die I want you to bury me here and not in Greenland."* He dies and his crew buries him with crosses at both his head and his feet so they name this place "Crossannes."

The crew then takes ship and returns to Greenland and on the way make only one definite waystop. Two are described, but one of these is told with such mystical events surrounding that it cannot be used for detailing the voyage. The definite one is said to be a river which flowed from the east to west and the ship is moored along the north (var. south) bank.

Some of the mystical events which surround both this and the following brief saga of Thorstein Erickson are interesting in themselves and may have some basis in fact. Somewhere along the homeward trip, a landing is made either by Thorvald, or possibly Thorstein, and two native children are captured and taken home. These two are well treated and survive to adulthood and so learn to speak Norse. They then tell of another land near where they came from where the people were white like the Norse and carried white banners before them as they "screamed." The king of this other land was named Valdidida.

The Saga of Thorstein Eiricksson

Thorstein Erickson is a brother of Leif. It is unknown if he was younger or older, but Leif seems to have been heir to Erick the Red. This saga contributes little to knowledge of Vinland, or even the courses thereto. As far as is known, he may never have even traveled there although it seems the more likely event that he might have accompanied one or both of the journeys of his brothers. Manpower, after all, was not so plentiful in Greenland in this period about 25 years from the time of Erick's settlement. Moreover, the tale is indistinct, confused, infused with mystery and mysticism. Many are the segments, so similar, yet often with different names that not much definite can be made of Thorstein. But he was a real person who seems to have lived to an age where he was seen as an accomplished seaman and independent personage. His sole contribution is to have been alive at a certain time, and to die at another.

The tale is of only slight relevance to the whereabouts of Vinland, but makes necessary contributions to the story by establishing certain durations, for the most tangible legacy he left to the epic is the person of his widow, one Gudrid Thorbjornsdottir. She was to become one of the most prominent individuals of a following saga, indeed a most remarkable woman and signal personality. She, whom the sagas universally praise for beauty, wisdom and sagacity becomes, at Vinland, the mother of the infant Snorri—first European child born on the North American continent. Almost without doubt, this woman of exalted character, who in later years actually made pilgrimage to Rome, is one of the original narrators of places and events of Vinland.

It is a disappointment, but the first we hear of Thorstein, is when he dies. His death apparently is given in three different and distinct forms. It is these differences which are so pertinent to establish timing of the following sagas, for the period of Gudrid's widowhood is crucial in settling the spans of residence at Leifsbudir. Fortunately, this time period has little bearing on the story as a whole, nor in establishing the whereabouts of the destination.

As he is an Erickson, we ought to give Thorstein more than passing attention, but space allows and requires only the following:

The most dramatic tale of his death is that it is said that when he hears of Thorvald's death at Crossannes, after the surviving crew returns to Bratahlid, he becomes remorseful and despondent; vows to go to Crossannes and recover his brother's body. He sets off and is either successful or not—it is not clear—but perishes in some manner on the trip. Possibly by the same natives who killed Thorvald, possibly by a "uniped" on the trip home, or by some other means. If this were so, then his ship would likely return the same year and Gudrid would be free only possibly in a short time.

A second version gives Thorstein heroic stature. On an unrelated expedition, his ship (his own) becomes becalmed in a "sea of worms." The ship becomes weakened from attacks of these worms and eventually founders and sinks. The "afterboat" is covered with a worm-resistant material and survives but there is not enough room in it for all. The crew draw lots for places in the boat and Thorstein wins a place. However, he has as part of the crew a young man who has been placed under his guardianship. This young man reminds Thorstein of his obligation to protect him as he has promised, and at this Thorstein replies, *"This is true what you say and I had not been of a mind to it. But I can see that you yearn to live, so I will stay with the ship and you will go in the after-boat."* If this version is true then Gudrid might have had a considerable time to wait for news of the lost ship—possibly two years, maybe more.

The third version is the one that sounds the more likely. It is the most interesting because of the extent of psychological symbolism. This version simply has Thorstein dying of famine and sickness up at the "Western Settlement" with Gudrid in attendance. On his deathbed, he makes certain prophecies, which in the event prove accurate. He predicts that Gudrid will survive and remarry well and go on to greater things. Then he dies.

It seems there is another couple there in the village, considerably older, and the woman of this couple also dies. The husband's name also happens to be

Thorstein ("the Dark") and he and Gudrid make common cause to survive, do so, and return to Brattahlid where he receives much honor and reward. Gudrid thus becomes Leif Erickson's ward but may not have remained so for very long. One senses that the people at the Western Settlement were not very unified through their mutual ordeal of famine.

The tenor is that the older Thorstein and Gudrid had no contact with anyone for several months although they seem to have lived in a village. Perhaps this reaction was one made in response to pestilences of yore, when people may have sensed that contact with others may have had something to do with transmission of disease. Typhus and cholera are generally associated with famines, and the great "black plague" of several hundred years later always had lesser precursors.

Thorstein's death is rather well developed, the tale holding that several times while he was near death he came to life, rising to sitting position, badly frightening the surviving Thorstein and Gudrid. When he did this several times, the older Thorstein held him to the bed again until he was dead.

At any rate, we now have seen two expeditions to Leifsbudir and conjunctures through Gudrid proceed toward that most important, most informative of sagas of Gudrid and her second husband Thorfinn Karlseffni.

The Settlement Expedition

Thorfinn Karlseffni and Gudrid Thorbjornsdottir

In this saga, we again meet people who become real personages to us. While the charismatic and possibly vainglorious Thorfinn Karlseffni is the more dominant person, his wife Gudrid, whom we met just previously as Thorstein's widow, is a major figure in our story. Almost certainly the tales she transmitted to her grandchildren are what we read today in portions of this saga, and she was, as I try to show in her short biography, quite a sailor. There is a hint in some of the tales that she may have been a survivor of the shipwreck that Leif discovered. It could be so but is not relevant to Vinland.

At some indeterminate time, she becomes enamored of Thorfinn. It is not known how long they may have known each other as the hazy background of Karlseffni gives some cause for speculation that he may have been resident at Greenland: and perhaps Brattahlid for some time. Indeed the sagas do not even tell us his legal name for "Karlseffni" is a colloquial nickname meaning essentially, "some hunk of a man." (It was actually Thordarsson. He came from the Orkneys.)

He asks Gudrid for her hand in marriage and she demurely suggests he make the request to Leif. As we have seen, as former sister-in-law and a dependent, her status likely was as some sort of vassal or chattel to Leif. The great explorer and leader set no obstacles in their way and they marry after a very short betrothal—*"the same winter."*

They set out to form an expedition to Vinland and approach Leif with the request to buy his "house" (this would imply the entire "vik"—settlement). Leif refuses to sell but gives free usage of it for as long as they wish, so the expedition is made ready.

From all indications, a sizeable complement was raised for the express purpose of settling Vinland, and the base complement was formed from populations of both Greenland and Iceland. It is difficult to establish the number of ships involved. Some translators give it as two, others three and some five. But from what we will see below, there were more than several years duration of the drama some of the ships must have made the journey several times and were probably joined later by others from both Greenland and Iceland. However, the narrator of the sea road was in one ship only and it is the impression of this reader that this particular narrator was someone possibly young and probably inexperienced with both the sea and the previous several voyages to Vinland. He or she uses some unfortunate phraseology, which has led certain strict scholastic disciplines to refuse acceptance, that the expedition even reached the coast of Vinland.

Recall, through this, that the tale of the journey out results from one narration but the drama, which increases in complexity, must be the result of numerous contributors. We follow only one of the several ships on the long, but by this time, well-known, sea road.

Alas, confusion sets in immediately for this ship is said to have departed toward the north and made an initial landfall at an island of which the narrator says, "... and they named this 'Bear Island'."

They proceed and come to the next landfall which has mountains and again the narrator says, "... and they named this 'Helluland'." This manner of speaking leads one to believe that perhaps it may not have been the same Helluland of Leif Erickson; so the sagas read.

Their next landmark is a place of which the narrator says, "... and they named this 'Markland'." They go on to the next which the narrator describes almost identically with an earlier traveler, "... and they sailed along some far-along beaches' which they named 'Wunderstrands' because it took so long to sail along them."

Then, as they are sailing along, they come to a place where they see the keel of a ship upon the beach. Some of the crew row ashore to examine this and return to confirm that it truly is a keel. (Some translators go so far as to identify this as Thorvald's keel, but even without this we can well suppose that it almost had to be. Moreover, for those unfamiliar with the visual aspects of observations at sea, we might presume that the lookout of the ship was actually expecting to see it.)

Many scholars, unfamiliar with this detailed episode, hold that Keelness was so named because it "resembled a ship's keel."

Here we come to the point of major divergence in interpreting the sagas. As mentioned above, certain over-precise scholars believe the expedition never even reached Vinland, although this is a difficult position to take considering the landfall of Keelness. The next major body of opinion holds that the ships actually reached Leifsbudir with no further difficulty. This is what *Flateybok* says. The third, inclusive of the entire episode of the next waystop, is the one which is supplemental to this story. It is often omitted entirely as having too much and/or too awkward detail to fit! There seems to be simply too much to it to discard the tale altogether as the many other analysts have done; and the detail, far from being superfluous, holds key elements to discovery.)

Vinland Landing

This landfall of Keelness is to become a major factor in establishing certain relationships to the ultimate destination that we are seeking. But for now, the ship(s) make off and come next to an island which, after some explorations, is named "Straumney;" "straum" meaning "stream," which is a reference to exceedingly strong currents around the island. (This is clearly stated by the narrator.) In the sagas as related about this place, there are cross references to "Straumfjord;" there apparently develops some traffic between the island and a fjord, although the landing at the mainland is never described. Events are described sometimes as at the island and sometimes at the fjord; so it is apparent that the island is in or adjacent to a fjord.

"Now, it happened that they had along two who were slaves to Leif Erickson and they had been given to him by King Olaf of Norway with the remark that if ever Leif had cause to need swift runners, these two should be used. These were a man and a woman named Haki and Hekja. They were Celts (var. Gaels) and their garment was a simple cloak fastened between their legs and open down the sides. And they wore nothing else. Karlseffni instructed them to go off south and explore the country down there so that he might know if it was good land or not. They set off and were gone for three days and returned; one was carrying grapes (var. grape leaves, vines), and the other carried a sheaf of self sown wheat."

The island is apparently a rather attractive haven and the sagas say that the entire group spent the rest of the season of good weather in explorations with the result that they did not attend to food gathering as they should. Consequently, during the following winter, a famine developed.

> *One day one of their number was missed and this was one Thorhall, called 'the Hunter.' He was a man of irritable temperament but was highly regarded nonetheless. He had been gone for three days when a party set out to find him. After a long search, they found him lying down upon a cliff, staring at the sky, mumbling and moaning and all the while scratching himself. They asked him what was the matter and he replied that "... it was none of their affair." They tried to talk to him but he would not reply. After a while, they asked him to come home with them and he said "all right" and came away with them.*

As the season progressed, it transpired that the winter was more severe than they had expected and their previous neglect at setting aside stores resulted in starvation. One day they came upon a dead whale stranded on the beach. This whale was of a type that was strange to them. Even Karlseffni, who was an expert on whales, could not say what kind it was. They cut it up and ate some but with the result that all who ate of it became sick. They then threw the carcass away (var. off a cliff).

There are said to be many birds on either this island or an outlying smaller Island they name "Birdsey." (Unclear translation but possibly meaning that the birds' nests are so close together that it is hard to step between them.)

In early spring, preparations are made for the coming year and then a dispute arises between Karlseffni and the same Thorhall the Hunter who had previously been lost. It seems that Thorhall wants to return to Greenland, or at least travel in that direction as, he says, *"... Leifsbudir lies in that direction, back beyond the cape."* Thorfinn and the bulk of the party hold that they wish to sail onward, *"... further south along the coast."*

Fortunately for us, this Thorhall considered himself quite a poet and as part of his argument he composed two refrains in an attempt to induce others to accompany him on his homeward course. Here they are in two versions selected from several extant translations:

> *Now let the vessel plow the main*
> *to Greenland and our friends again!*
> *and leave this strenuous host*
> *who praise this god-forsaken coast!*
> *To linger in a desert land*
> *and boil their whale by Furdirstrand!*

It is not widely known that there was actually a short form of literary effort on the coast of Vinland but there it is. Since it yields us a most important clue which will be investigated later, we will examine another translation which shows the extreme variations possible in reading the runes; same refrain as above.

> *Comrades let us now be faring*
> *homeward to our own again!*
> *Let us try the sea steeds daring*
> *Give the chaffing courser rein!*
> *Those who will may bide in quiet*
> *feasting on a whale steak diet*
> *in their home by Wunderstrand!*

("Furdirstrand" is the same as "Wunderstrand," meaning long beaches or "marvelous beaches.")

Thorhall's Departure

Thorhall the Hunter is unsuccessful in swinging others to his views. When it is time to sail (he evidently had his own ship) he splits off from the others and goes on his own way with an apparent "bare bones" crew of nine men. He leaves with this parting shot of embittered verse:

> *They flattered my confiding ear,*
> *with tales of drink abounding here;*
> *my curse upon this thirsty land!*
> *A warrior trained to bear a brand,*
> *a pail instead I have to bring,*
> *and bow my back beside the spring!*
> *For ne'er a single draught of wine*
> *has passed these parched lips of mine.*

Here is a variation for added interest:

> *When I came, these brave men told me*
> *Here the best of drink I'd get!*
> *Now with water pail behold me,*
> *Wine and I are strangers yet!*
> *Stooping at the spring, I've tasted*
> *all the wine this land affords.*
> *Of its vaunted charms divested*
> *poor indeed are its rewards!*

The affection in which he seems to have been held becomes more apparent as someone keeps tabs on him, avoiding abrupt eviction from the sagas. His courses are unfruitful and also unlucky. He suffers a terrible trip and eventually makes landfall off Ireland. His landing there is among a people long exasperated with incursions of pestiferous Vikings. Thorhall and company are imprisoned and enslaved, beaten and abused; it is said that Thorhall was killed there eventually. News of this unhappy result reached Greenland many years later.

(This failed trip also indicates the great distance of Vinland from Greenland as well as the location of Leifsbudir. For we can now presume that Leifsbudir must lie somewhere beyond Straumney, which is beyond Keelness, which is beyond Crossannes. And Crossannes, it may be recalled, seems apparently to be not altogether distant from Leifsbudir.)

The main story goes on. There is a minor version, which states that Karlseffni went on with only forty men for a period of two months, leaving one hundred on the island, and then returned to Straumney after certain events. The most acceptable version says that all, or most, of the complement pushed on and *"... after a long way"* (var. "time") reached a place which they named "Hop."

Hop becomes a sizeable settlement either immediately or soon thereafter. It is the most completely described of the Vinland settlements and the population at one time was as high as 150. Some scholars believe that this figure may be over 200. Moreover, they are known to have had cattle and other livestock along with them. A rough description of the settlement goes, *"... there were houses (shelters) down near the water and more farms further back on the other side of the hill," "... all the hillsides were covered with vines and all the valleys grew self sown wheat," "... the river flowed down the land (from the north) into a lake, and then into the sea,"* and *"... halibut were caught in pits dug in the marshes."*

(It is the contention of this work that Hop was, in fact one and the same place as Leifsbudir. Few scholars have made this connection, but when the sagas are read in their entirety this becomes more evident.)

Events proceed in the following manner, however. At some time (var. *"following spring"*), contact is made with a group of natives who are described, always with the derogatory name of "skraelings," which means "screechers" or "barbarians." They are described as *"small, dark ugly men with unkempt hair, broad cheekbones and bulging eyes."* This initial contact, though cautious, is fair enough and soon trading begins between them. It is in this period when the variations—not severe—become apparent between the Greenlandic sagas and the Icelandic ones.

One senses at all times the two groups and also the several individual narrators who lived out the drama. One says the natives *"... came out of the woods,"* the other says they came by "boats." Of the water approach it is said that a number approach, stand "*off the point*" and wave their "staves" (var. shields) "sunwise," (i.e., east to west); this is interpreted as indicating peaceful intentions. The Norse hold out a white shield as an indicator of peaceful intent. This is interpreted by the natives as the Vikings wish. The natives land and on this and several subsequent visits, trading is commenced.

The natives have "dark" or "gray" pelts, the Greenlanders red cloth and milk. The sagas give good impressions that the Norse think they have the better of the bargain, for they say that they cut the red cloth into strips and when they start to run short they cut the strips narrower, but the natives pay as much or more as for the wide strips. It is also said that the natives are very enthusiastic at sight of the milk "in pails" and wanted that and nothing else, "... *they left their goods behind and carried their purchase away in their stomachs.*" The natives also go to great lengths to buy some of the Norse iron weapons, but Karlseffni forbids this.

During one of these trading visits, "... *a bull belonging to Karlseffni*" suddenly gave a great bellowing and "*—charged down out of the forest*" onto the point where the trading is taking place. The natives take fright and escape into their boats while the Vikings subdue and placate the bull. This done they turn and try to induce the natives to again come ashore but this they will not do, only stare sullenly, and then row off "... *south beyond the point.*"

Karlseffni recognizes that there is now some danger from these natives and he orders a palisade to be built for defense. He also devises a battle strategy of which several versions—all bad—appear in the sagas. One of these is that a battleground is selected "... *between the lake and the river"* to where they will proceed "... *when the battle commences, driving their bull before them.*"

This battle does ensue; and provides a wealth of vital insights to the place. These will be addressed shortly. In the event, the Vikings do not do as well as we might have supposed, and are driven to, or make, a strategic retreat. They lose two of their number; from the way the sagas are expressed, they are not of the feeling that they must escape the settlement precipitously, but later, after an indeterminate time and much debate, they decide—or Karlseffni decrees—that there will always be trouble between themselves and the natives. So they decide to leave the area and then return to Straumney, the island where they had suffered famine previously.

This second settlement of the island is more successful. They remain there for some time. The total time from the stay at Hop to the time when Thorfinn and Gudrid leave the coast to return to Greenland can be established very closely, for their child, Snorri (known to have been tended as an infant during the battle at Hop) is said to have been three years old when he left the coast of Vinland.

The Skirmishes at the Settlement

One of the most informative items that can be studied is the tale of the battle that occurred at Hop. Nobody really seems to like war—at least after it has started—but much of human progress and much of human knowledge has been in response to the need to document the misfortunes that result. This is the case with this battle; it yielded a plethora of clues for me to follow in my research.

Fortuitously, the story comes to us through the two sources, the Icelandic and the Greenlandic. The two descriptions are distinctive and unique but qualified by the difficulties of translation. Leif had apparently become aware of natives living in the area, as witnessed by what may have been a martial order in organizing his crew in the manner he did. As events befell, it seems that he did not have any direct contact with these natives. His omission of direct remark of them is less noteworthy than if he had stated an absence of aborigines, since we know that all lands of the world—especially all sizeable lands—were inhabited by that time.

Essentially the scenario of the battle is as follows: The natives, who apparently reside at some distance, make a sudden dramatic reappearance in their boats and an action commences upon the point whereon the trading had taken place previously.

Evidently overwhelmed, fearful of the event, or for strategic reasons, the Norse make a retreat to palisaded structures. Fighting goes on there for a time and the natives hurl a device over the wall, which makes such a *"hideous noise"* that a panic results, and a second retreat commences northward, *"… toward some cliffs that they knew."*

Whether the defensive position, which was the destination, was ever reached is not clear, for the fighting seems to progressively wind down. The victorious natives break off the battle, return to their boats and *"… row off south around the point."* That is the sum of the battle but it may have been more complicated than it appears.

Detail

As we have seen in the previous section, relations with the natives seem to have started at least reasonably amicably. The Northmen met and traded with them apparently on several occasions. We can see quite clearly the Norse point of view of these natives, not only in the derogatory expression used—"skraelings"—but also in the sly inference that they felt they had gotten the better of the others in the bargaining. From the native point of view, there must have been abundant curiosity about these people who had now appeared from the sea on three separated occasions and showed every indication of interest in remaining at the spot. Their initial approach seems one of investigative but wary interest, since they came in numbers and "stood off" until a "meeting of the minds" was established.

It is interesting to speculate that perhaps they, too, felt they had the best of the bargain as it seems likely that the pelts traded were rabbit, game of great abundance in the neighborhood, and a notorious "glut on the market" in fur trading all over North America.

But after the charge by Karlseffni's bull and, no doubt, with the dubious novelty of interesting strangers wearing off, the Indians take themselves away and are seen no more for a period of three weeks. Karlseffni seems to have had some premonition of impending trouble for he takes certain actions intended to bolster defense of the place, including the building of a palisade.

There is a divergence in the details of the battle, but they essentially conform with each other in major factors. It is evident that the battle took place in the springtime; it is so stated and also that one of the items of barter was *"milk in pails."* But the manner in which the skirmish started is interesting for it is given in several ways, three of which are all plausible, are all possible, and all give insights as to how disparate individuals entered into the fray.

It is said that one morning they looked out over the lake and saw a multitude of native boats approaching; *"… they looked out and saw boats, they came as a torrent, appearing as charcoal strewn upon the water."* (This perspective was a "key" clue.) Again the natives stay somewhat offshore but this time wave their "staves" (var. paddles) anti-sunwise, i.e. unnatural direction; from west to east, and at the same time they make a noise, "... *as of threshing*" (repetitive thumping). Their intentions are read by the Vikings as being hostile and a red shield is held forth as token of accepting battle. The natives either force or are allowed a landing for fighting which immediately commences.

One of Karlseffni's narrators gives him credit for establishing a strategy of sorts during the peaceful interim. Since the strategy is inherently faulty and does not even have the advantage of ultimate victory it seems doubtful but might well be true. In this, Karlseffni locates a place in a clearing "... *between the lake and the river"* and he says that at the appropriate time they will "... *make for this clearing, driving our bull before us."* In the event the sagas say the natives came right to the place as Thorfinn had planned.

Another of Karlseffni's strategies is defined as having a small force hold and delay at the point while others prepare. This coincides with what is known in the first version and may be likely. This consideration of marshalling of manpower should be considered in viewing the somewhat less than spectacular result of these armored men of a warrior race. Without an alarm, it would take at least a half-hour for several runners to do the job, and the distance from the portage to the point is only a matter of few minutes paddling if pressed. Surprise and psychological unpreparedness must have had a great effect on things.

Gudrid's View

Another variance from the Icelandic sagas—apparently from Gudrid herself—says that a native woman appeared at the house where Gudrid was tending Snorri and a brief exchange was interrupted by a *"great crash"* when a native male was slain for attempting to seize a Viking sword or ax. The exchange goes like this:

> Gudrid was sitting with Snorri who was asleep in his cradle when she looked up and saw framed in the doorway a Skraeling woman. She was of small stature, and had a band around her head of auburn hair. She wore a black tunic and had the largest eyes ever seen in a human head.
>
> She said to Gudrid, "What is thy name?"
>
> Gudrid replied, "My name is Gudrid. What is thy name?" At the same time, she motioned for her visitor to sit beside her.
>
> The woman moves as if to comply (and may even have done so) when they are startled by the loud crash, and the woman disappears. The crash is attributed to an Indian attempting to seize a Viking weapon, at which he was slain and, in falling, knocked down some barrels or casks.

If this is true, then these two Indians must have been inside the palisade—and probably peaceably—when this happened.

Retreat

The fighting on the point goes badly for the Vikings and they are forced to retreat to the palisaded buildings. Some versions state that this was a tactic ordered by Karlseffni, but if so, was a very faulty one, if we can believe the common but possibly exaggerated claim of excited combatants that they were slaying many enemies.

The natives in their turn now spring a little surprise of their own. They have a device described in the sagas as *"... blue-black and about the size of a sheep's belly"* which they slung from a pole over the palisade where it lands *"... with a hideous noise."* Whatever this was, it creates a panic and another retreat starts. This retreat is specified to have headed north along the river to *"... some cliffs that they knew."* They are harassed by the natives and soon pass Freydis' house, within which that irascible woman was apparently in ignorance of the goings-on. She comes out and rebukes the men strenuously.

Her remarks cannot be omitted without losing something valuable from the tale:

"Why do you run from men such as these? You should be able to slay them like cattle. Why, if I had a sword I could do it myself."

Unarmed, she is left, but sets out and attempts to keep up with the retreating party. She is "big with child" and "much slowed" and is unsuccessful in this and is about to be caught up when she comes upon the body of one Thorbrand Snorrason who has fallen with a flat stone penetrating his skull. His sword still lies beside him and Freydis snatches this up, bares her breasts and stokes the sword on them; which action dismays the natives who break off combat, return to their boats and *"... row off south around the point."*

The native leader is described. He is said to be taller, lighter, and handsomer than the others, *"... a real leader."* The natives all along had been trying to obtain some of the Norse iron weapons (even during the trading) and at the conclusion of the battle the "skraelings" unhurriedly take time to examine one, which an individual brings to his chief. This leader tests this—either a sword or an ax—by striking a stone, which breaks it (the weapon, in one version), or by striking a companion, which kills him (the companion, in the other). He expresses disgust—one can almost see him, the sagas are so lucid—and throws it with all his might "... *out into the water.*" After the battle, the Vikings again regroup and resettle and discuss the battle.

They claim to have inflicted many casualties while suffering only two men dead themselves. Most significantly, however, they also debate a matter which enabled me to extract another clue. Someone apparently has made the claim that a party of natives came by land—from the west, a separate attack from that of the boats. This claim is discounted. The party concludes (or Karlseffni decrees) that the landward attack was "... *an illusion.*" For some reason, they cannot believe that anyone approached the settlement from the west. Moreover, Gudrid's claim to have spoken with the native woman just as the battle commenced seems to have been accepted only reluctantly and because she had a reputation for truth. The sagas say with an atmosphere of doubt that "... *only Gudrid saw her.*"

At some later time, the impression is that it was not immediate, Karlseffni decrees that they will not ever be able to remain on anything like amicable terms with the natives. As a result, the party decides to return to the island Straumney. During the first year at Hop, however, they had cut timber for cargo and "*set it out on a rock to season.*"

They may have taken this timber back to Greenland with them, for Karlseffni subsequently made a trading voyage back to Europe. It was said that this cargo was the richest ever to have sailed from Greenland.

In Europe, someone wished to purchase the figurehead of his ship and Karlseffni refused all offers, saying that the figurehead had been made of "mossur, from Vinland." Most scholars believe that "mossur" is Maple or possibly curly Maple, perhaps a burl of a Maple tree, which is ideal material for carving of items like figureheads.

Scholars and archaeologists also believe that Karlseffni, Gudrid, son Snorri and possibly another son (Bjorne) settled in the northern reaches of Greenland at a place named "Sandness" where they resided for some time before eventually moving back to Iceland. Their Icelandic farm is also well known, named, and has been archaeologically excavated. These "northern reaches" of Greenland might include the ill-fated "Western Settlement" of Vikings. It may be recalled that Gudrid's first husband, Thorstein, seems to have died there with Gudrid in attendance. Thus, she would own property at that locale.

The Saga of Freydis Ericksdottir

From across the time and distance to that long neglected Viking colony of Greenland, it is rather hard to grasp that entire generations of people lived out their lives in relative peace in a European manner on what we view as a desolate island: cold, bleak, harsh Greenland. But we also know now its climate was in that period somewhat milder than today. A whole separate culture existed there, somewhat remote from the parent culture of Iceland and even capable of sending out explorations and collective expeditions.

Such literature as survived from Greenland were documents which went back to Iceland with people who returned there, persons who removed for whatever reasons including marriage, and sometimes appeared in the Icelandic and Norwegian sagas in indirect, as well as direct, references to the land to the west. Leif clearly had troubles mixed with his good luck and among others was an illegitimate half-sister named Freydis who is one of the most signal personages of the Vinland voyages, because it is known that she made the trip to Vinland at least twice.

Her personality comes down through the ages in the epics as hardy, effective, shrewish and venomous. However, in their youth there is evidence that Leif thought quite highly of her—she may have been older than he; or just as likely, considerably younger. She had a number of children, one of whom was apparently born at Leif's house in Vinland. She first appears as a participant in a battle, which took place at a place called Hop, which is a Norse version of the English "Hope," meaning a marshy tidal inlet or haven of possibly dangerous approach.

The settlement there was sizeable and populated by as many as 165 persons. Primarily amicable contacts with natives deteriorated and terminated with some natives being charged by a Viking bull. The visiting natives ("skraelings") depart with sullen mien and apparent demonstrations of malice, return in several weeks' time with many reinforcements, and commence a battle in which they ultimately are successful in inducing the Vikings into an uncharacteristic retreat.

The retreat passes the house wherein Freydis resides and she, perhaps previously unaware of the fighting, attempts to rally the men by ridiculing them. This does not work and she is forced to join them but is "... *much slowed, as she was big with child.*" The retreat leaves her behind and she is about to be caught up by the natives when she comes upon a dead Norseman. He is named and described as being felled by a flat stone, which protrudes from his skull. His sword is beside him, and Freydis snatches this up, brandishes it, exposes her breasts and strokes the sword on them. The dismayed natives then retreat in turn, or simply break off the battle—the sagas do not evidence haste—and they depart in their boats. Formidable woman!

Hop is later apparently abandoned because of the precarious position or proximity to the disturbed natives, and the band removes back to the island of Straumney. Freydis and others reappear back in Greenland some few—at least three—years later.

Freydis decides to found an expedition of her own to Vinland and forms a partnership with two brothers named Helgi and Finnboggi. They travel in two ships which make the remarkable and noteworthy accomplishment of arriving at Leif's houses a fortnight apart, not having sailed in—or at least early parting—company. The brothers have arrived first and ensconced themselves in Leif's house. When Freydis arrives, she takes affront and evicts them with considerable asperity. They go off and take up residence in a single large structure. Their numbers are stated as being twenty men and five women. At departure, they had agreed to maintain equal crews, but Freydis surreptitiously secreted five extra men who were discovered by the others only at arrival. The complement here at this time is, therefore, 55 persons, of whom 6, likely more, were female.

Initially, the two crews enjoy amicable relations, engaging in sports and games. But soon the two parties become antagonistic because of the enmity between Freydis and the brothers, and the seasons and winter proceed with growing bad feeling between them. Then occurred an incident which in horror and deceit lives in the annals of the north and contributes to the ultimate abandonment and disfavor of Leif's "budir," and indeed, disturbs Scandinavians to this day. Most of the story must have come from the lips of Freydis herself.

One night she arises while all are sleeping and goes over to the other house of the brothers. She does not enter but stands quietly in the doorway and is eventually seen there by the half-wakened Finnboggi. He asks her what she wants and she replies that she wants to talk to him and desires him to come outside for this. He arises and goes out with her, and they sit on a log just outside and talk, apparently a low, subdued conversation. However, we have only Freydis' word for it, for Finnboggi did not survive the sunrise. Freydis begins by asking him how he likes the place, and he replies that he likes it well enough but is disturbed by the dissension between them. She claims to be as disturbed as he, and then goes on to say that her purpose in coming was to propose a trade in boats as hers is smaller than the brothers'. She later claims that he agrees to this trade and after a few more remarks, they part. She returns home (to Leif's own house) and Finnboggi goes back to bed.

When she is home, she returns to her bed and awakens her husband, Thorvard, with her cold feet. Perhaps he was not asleep after all for he questions her on where she has been. Rather than tell him the above, she says she has gone to the brothers to trade boats and they have insulted and abused her, and that she expects Thorvard to do something about it. She berates him unmercifully—it must have been a long night for him—and she eventually threatens to divorce him on their return to Greenland.

Goaded thus, he takes his crew and removes to the other dwelling and surrounds the sleeping crew. A party enters and binds the men, 20 in number, and then they are brought out one by one and killed. The five women are then brought out but there is universal refusal to kill them, at which Freydis *"... takes an ax"* and *"... leaves them dead."*

She swears the crew to secrecy under pain of death and eventually the party returns to Greenland—some say in the brothers' larger ship. The saga called *Flateybok* is explicit that this homeward trip commenced, *"... in early Spring, and they arrived in Greenland after a happy trip in early Summer."* (This key statement has been completely overlooked by most as to its significance. It means a journey of at least two months at the very least, and if we consider that the seasons are later at northern latitudes it could well represent four months travel time. It certainly places Leifsbudir a very long distance from Greenland.)

Soon rumors abound as to the fate of the missing crew. Dark suspicions arise as to events at Vinland. Leif, who had inherited leadership on the death of Erick the same year as his return, has two or three of the crew seized and put to torture. They separately tell identical stories which are accepted as thus being true.

Freydis is brought to trial and convicted but rather than mete out severe punishment, Leif simply makes the anguished statement, *"I have not a mind to do unto Freydis what justice demands. I only say that her and her clan and children will amount to but little."* This explicit and probably most effective curse is the final direct reference to Leifsbudir in the famous Vinland sagas. There is no further remark or mention in the Greenland epics of any interest in the place. It is possible that these devout Christians felt that the settlement to which they had traveled for over twenty years was now unfit for further habitation.

On the whole, it is a dramatic and effective tale, as are many others surrounding the Vinland voyages. One can almost feel the wry, dry menace of Freydis' madness. The scene where the two antagonists meet and sit on the log for talk can almost be brought to the mind's eye.

Conclusion

The story has endured for over 960 years through interpretations from folklore, memory, runic script, Latin, and Icelandic to English. Through all that time, from all that distance, over so many peoples' tongues from that fey settlement of human daring, the actual vision of happenings has been transmitted to us. The lost ruins of the old Viking settlements on Greenland came to light and awakened, sometime in the 16th century, interest in the old explorations of many years ago. A host of individuals and some companies, perhaps even the great Admiral Columbus, upon hearing and attempting decipherment of these even then ancient sagas, have made strenuous efforts at finding the precise locale of Leifsbudir—that river, lake and terrain whereon the son of Erick the Red had built his houses.

The literature on the subject is voluminous, filing whole cases in libraries, the theories abounding, replica voyages made, seminars, college courses, theses, theories, articles, books, pamphlets, speeches, talks, and arguments. All have attempted to fulfill a void of mystery felt by generations of historians and Scandinavians for whom the old tales of the hero and the voyages thrill the hearts and resonate in the intellects of humankind.

So far, none of these studies has born the fruit of acceptance by all persons, or even a substantial portion of them. The site of L'Anse aux Meadows in Newfoundland has so far come the closest to acceptance by being demonstrably Norse, but as time goes on and comparisons are made against the sagas, it too, becomes embroiled by doubts that it is Leifsbudir. Certainly, that Canadian locale will not bear attempted duplication of navigational detail for it is in a straight line to and from Brattahlid in Greenland and would entail only days' travel thereto, not the long and elaborate voyages we can particularize from these epics. Therefore, the search has lasted for over four hundred years.

I have reason to believe at this time that the search may be successfully concluded. A site which matches in all specifics and in all respects concerning the terrain and the traverse has been found and thereon exists evidence that at one time a people very like the Norse resided there. Perhaps, just perhaps, we have found the place where Leif Erickson and his hardy band of thirty-five made their epic landfall a thousand years ago.

Note: If any text in the world has entered the public domain, this one of 800 years of age certainly must qualify. The author requests that any direct quotations from this work be attributed to the writer as a courtesy.

The reader might note the difference in spelling of names at partition of this work into chapters. This is done simply for reasons of convenience to English speakers whose train of thought might be distracted by the apparent archaic form of names. In fact, it is a highly complicated issue in that the name "Erick" is not a name at all, but a title somewhat of the level of an English "Earl." The man we know as "Erick the Red" was actually named Thorvald, and was known by the more popular form. That was a Scandinavian custom for the leader of a settlement of an extended family inclusive of distant relatives, hirelings, slaves and their offspring sometimes to the size of a village. Therefore the name of the man we know as Leif Erickson is, in Scandinavian usage, "Leifur Eiricksson", "son of Thorvald, the Erick of Bratahlid known as "the Red." But his, as well as his siblings, were actually, in law, named Thorvaldsson. Still, even though he is universally known as Erickson, we English speakers take this liberty and we should honor the name in his own language, "Eiricksson" derived from two words, "Eirick's Son." It has two "S"s without the apostrophe. So our dividers above pay honor to the man and his culture, but we think the narration smoother with his more universally accepted name in the language of this book.

The narration of the sagas is drawn from an extensive reading program that now exceeds some 33 years. At this time, to my belief, it is nowhere else to be found to anywhere near this detail in the English language. The reader is cautioned that the essence of the tale has been not only recorded late from Icelandic chronicles and legends, it has been translated by varied scholars with as many viewpoints.

We seek the truth of the story but this span of time and travel of the written word makes us always suspicious of what it was that actually happened a millenium ago. But we are assured that *something* happened and this is the only source to these episodes in history. Now published and extant in Iceland itself for some 25 years, there may be an expected amount of dispute, argument, and even dissension, but these have not appeared.

These tales, as I understand them, led me to a place that coincides perfectly with what the sagas say, and then to an element of scientific testament, the epics would seem to be pretty close to the events of that olden time. This book demonstrates that there is a place exactly as the sagas describe. There actually are traces of the long ago incursion at this place. The Vinland voyagers did leave a legacy that we can detect. And from this we can conclude that Leif Erickson's settlement site is re-discovered.

Part III

▼

The Vinland Sagas as a Guide

An elusive historical "target," aimed for by many for centuries.
A "voyage" that commenced as a lark but developed into a serious endeavor.

The Voyage of Wave Cleaver

When I put these tales, sagas, and stories together, they formed a sequence of events about Vinland. (1) Bjarne Herjolfson saw the coastline of Vinland. (2) Leif bought Bjarne's boat and sailed there. (3) Leif's brother, Thorvald, used the same boat and sailed there. (4) Thorfinn Karlseffni negotiated to use Leif's settlement. (5) Leif's half-sister, Freydis, also negotiated to use Leif's site and, having been there before, recognized Leif's house upon arrival.

Why is this sequence so important? We can now accurately describe the coast of Vinland and its environs. This, together with the knowledge that Freydis' return voyage consumed about three months, indicates considerable distance and conjures up a cohesive new vision to help us locate Leifsbudir or Leif Erickson's original settlement.

I have postulated that Leif's camp and Karlseffni's camp (which the latter called "Hop") are the same place. Other studies hold that the two were so remote as to possibly be on separate continents. This came as a revelation as I considered the new research and sagas about the layout of the coast of Vinland. Thorvald was certainly at his brother Leif's house and he and Karlseffni shared a common landfall, which, therefore, must also be in Vinland. This landfall, "Keelness," where Thorvald scraped bottom and repaired their keel on a headland, became a marker of the courses. Karlseffni's similar experience suggests that he may have found Leifsbudir, the place he was aiming for.

I also noticed that Leif's half-sister (Freydis) was present on this expedition. Since she was easily able to locate the site years later to her half brother's house and camp, she must have been there before. Leif never named his camp, but Karlseffni did. Their landings at the river mouth were similar, unusual, and unique. The site I propose coincides with Leif's somewhat sparse and Karlseffni's quite detailed accounts.

We have for descriptions of Leifsbudir: a river mouth, river, lake, hills, and orientation of the settlement, placement of structures, zoology, botany, and distance inland from the sea. Fortunately, some archeological work has been done in the area of settlement that I am proposing. Several indications show activity well predating 1492 and a scientific survey of my proposed river and settlement actually showed a drastic change of activities at the time of the Vinland voyages according to archaeologists.

Up to now, the site at L'Anse aux Meadows has been accepted as proof of the contact of Vikings on American shores. However, it is now conceded that the landing was accidental, short lived, had no permanent effect, and seems to have been simply a way station. It was used as a place to build, repair, and maintain ships.

According to the sagas, Hop is described with some of the following criteria: It was a considerable distance away from Greenland. Leifsbudir itself was at least a three-month journey and according to conventional scholastic interpretation, Hop was further away than that.

Traveling away from Greenland, it was beyond a prominent north-pointing cape. Another saga described a landfall in the same way, and was probably the same one. As well, it was beyond a sizable island noted for strong currents and which was so large as to take at least three days to explore. Upon this island grew grapes and self-sown wheat. This was important because Greenland was too cold and too barren to cultivate grapes or wheat. The island settlement was also at some clime where a species of whale unknown in northern seas lived.

Straumney ("straum" means strong currents) was described in sagas. It could not have been a great distance from a mainland with the fjord they named Straumfjord, because in the first year traffic commenced between them suggesting a close geographic relationship.

Salmon were described in the sagas as being in the river of Hop. Most people agree that salmon's southern limit has always been the Hudson River at New York City, so they were north of that point. F. Donald Logan, *The Vikings in History*, stated

> At the present time, salmon are found no further south than 41° N and wild grapes no further north than 42° N. This would place Vinland somewhere between New York City and Boston.

Logan also noted that the climate in the eleventh century might have been slightly warmer allowing grapes to grow farther north.

Hop was located along a river that flowed from north to south and had some unusual configuration. The saga holds that "... the river flowed down the land into a lake, and then into the sea." Lakes as we know them—fresh water—are not common in that type of arrangement near river outlets. The Norse definition of "Hop" confirms this. This river could not be entered except at high tide, a factor that appears to delineate its proportions.

The Hop settlement that developed was near where halibut were caught, according to the sagas. Halibut do not enter fresh water but limit their inshore breeding to brackish tidal water only. Therefore, Hop was adjacent to oceanic waters.

The settlement was along a hillside on the side of the lake near some prominent cliffs, according to the sagas. The description stated, "... there were houses down near the water and more farms further back on the other side of the hill." The hill was relatively high and steep. The vividly described vista strongly confirms also that it faces east. This prospect overlooked a "point" (of the shoreline) with a general layout of north/south, and it was heavily forested.

Aborigines dwelt at some little distance inland from the site, said the sagas. There seems to have been some sort of geographical barrier between the native and Norse sites. Post-battle discussions by Karlseffni and some of his soldiers dismissed as unlikely (or "an illusion") that an attack had come from a westerly direction. The natives approached the "point" by boat from the south and departed toward whence they came.

There were at least two sizeable rocky outcrops somewhere nearby according to sagas. One was where timber for cargo was set out to "season." Another was a spectacular place eminently suited for defense. This spectacular place was described in the sagas as located to the north of the settlement and near the river. It was at some reasonable distance but not more than two or three miles at most. These "cliffs" were the key to my discovery because they might be permanent and recorded. They were well-suited for defense, since the Viking party, armed and armored, abandoned home and palisade in a panicked retreat toward them as a last ditch haven, according to their sagas.

We here deal with many factors. One is that a north-south layout seems indicated. I found such a layout. The overall situation is rare, especially with the hop-like lake and cliffs near a river. The site I found in Rhode Island fulfilled many of these requirements.

My original goal with *Wave Cleaver* began as a search for Karlseffni's Hop, rather than Leifsbudir. Translated, this means Leif's "budir" or "boothur" or portable home site office, shipboard residence, or campsite command post. The word "Leifsbudir" never occurred in the sagas. Only a few writers who referred to it used that word. The fact that Leifsbudir was never named or mentioned in the sagas will have great significance as we proceed.

Both Leif's landing site and Hop were said in the sagas to be located in a district where daylight hours were nearly equal to nighttime hours; where grapes and self-sown wheat thrived; and where the climate was mild enough for cattle to dwell outdoors all winter without fodder. It had to be a considerable distance from Greenland for one voyage, said to be a fair or happy one, took at least three months from Vinland to Greenland. This is a notable factor to contemplate concerning seamen, who could sail with regularity from Norway to Iceland in seven days and from Norway to Greenland in ten. In our day, we sail straight across the sea, of course, instead of hugging the coastline as the Vikings usually did.

Obviously, the sagas are not historical documents. They were stories based on oral history told for the edification of people interested in their past. As such, the mention of Eyktarstad and Dagmalastad (ancient Norse astronomical terms for sunrises and sunsets) may have been inserted by a writer to give interesting detail to his story. Some have argued that a storyteller would not know these details unless they came down to him from someone who had been there. I would contend that this is just the sort of detail a good storyteller provides. Additionally, a sea-faring culture such as that of Iceland would be aware that sunrise times change as you move further south.

Phenomenal findings of lost, forgotten, and overlooked references and records have uncovered pertinent facts. We now know that the environs of the site I've studied have been proposed as Vinland by the earliest explorers and settlers of the district. They had directly observed things that were later nearly forgotten but fortunately carefully recorded. Possible European contact was a live issue in olden days. Indeed, a visitation by some "advanced culture" is recorded in documents, which had traceable effect on the anthropology, legends, and even language of the natives of the region.

If imagery of events of a thousand years ago seems distant and ethereal, it would seem to be from force of habit, for the Vinland adventure is relatively recent and tangible. Erick the Red belongs to the ancients. He was a swashbuckling Viking to the core. His son, Leif, was a transition man, surprisingly a missionary with a gentle streak to his nature. Yet he seems spectral to most. I imagine this is because of the much later abandonment of the long-lived Greenland settlements, so neglected in American history.

Thorfinn Karlseffni, the notable merchant who married Leif's ward, was an essentially modern man. He is a noted figure in Iceland history; his family chronicle is well known. His origins in the Orkneys were recorded and his success in business enterprises remarked. His wife, Gudrid Karlseffni, is one of the most remarkable women in history, not simply because she mothered famed Snorri (often written as Snorre) in Vinland.

Many have said that the first white child born in America was Virginia Dare in the Roanoke Colony of Sir Walter Raleigh. Virginia was the first white child born of

English parents. Virginia was not even the first Christian child born in America because Gudrid was a devout and strong Christian. Snorri (Snorre) would go down in history a notable man who was also a strong Christian. He and his brother became officers of the Catholic Church in Iceland.

Gudrid was the first person recorded to have spoken to a New World Aborigine as well as face to face with Romans in their distant continental city. Obviously many others spoke to aborigines but their conversations were not recorded.

My analysis and discovery of Vinland and the sagas extended from 1977 to approximately 1985. I completed a narrative from the sagas and published it in 1985. I submitted it to the National Library of Iceland, the Government of Iceland through the office of the Ambassador in Washington, D.C., and the (then) President of Iceland, Mme. Vigdis Finnbogadottir. They were accepted and I was encouraged to proceed.

By the time I published my work, I considered my working hypothesis to be whether Pettaquamscutt in Rhode Island was the site of the Vinland described in the sagas. My basic argument was "if it walks like a duck, flies like a duck, sounds like a duck, and looks like a duck, then it probably is a duck." By that, I mean that the site correlates perfectly in all respects with such information of descriptions extracted from the sagas.

I consider my circumstantial case for the Vinland settlement very strong. I believe that I have identified the correct site of Hop at Narragansett, Rhode Island, a few miles north of Point Judith, just beyond the cleft of Narrow River where lies Pettaquamscutt Cove Lake, south of Pettaquamscutt River.

I believe that Pettaquamscutt fits the description of Vinland's location. Its terrain, climate, flora, fauna, and orientation compare perfectly, and all activities extracted from the sagas correlate reasonably with actions described in the narratives.

Years ago, when I commenced my search, many thought the Norse site at L'Anse aux Meadows in Newfoundland was the Leifsbudir (Leif Erickson's settlement) of the sagas. Most analysts felt that Karlseffni had journeyed further to the south and there founded two settlements, one on an isolated island called "Straumney," and another on a Native American-occupied mainland named "Hop," (pronounced "Hope).

I set myself the task of finding this "Hop," for there were many more clues in the sagas related to it than there were for Leifsbudir, and it seemed that it must certainly be nearly as important. Following a three-year reading program and two years of explorations, a site was located. From that date to this, my supporters and I have been attempting to determine conclusively whether this was, indeed, the correct site.

I propose that the site of Hop and Leifsbudir is at the outlet of Narrow River in the town of Narragansett, Rhode Island, where lies Pettaquamscutt Cove Lake. Bearing from it to the north is Pettaquamscutt River. I believe that the type of boat used by Leif Erickson and Thorfinn Karlseffni was a Gokstad ship or something very similar, according to the boats of that era found in Scandinavia.

A Gokstad ship was a Norse longship with 16 oars on a side and measured about 76 feet in length. We know of those ships because they often served as burial chambers when an important ship owner died. They were preserved under a barrow of impermeable blue clay. A Gokstad ship was usually constructed of oak to ensure strength. Few trees would have yielded straight timber much longer than 76 feet.

I created my real and imaginary boat or *Wave Cleaver* and launched it in 1975. I wanted to go and approach a site in a boat like those Norsemen who found their landing sites in America.

In general, from the sea, the entire coastline seems "low and wooded" as was described by Bjarne Herjolfson near 986 A.D., but it takes on more complexities as one approaches closer.

Outer (north jutting) Cape Cod is also a low-lying sandy feature with low forest covering it. It is a geographic feature resulting from sea currents, mainly the Gulf Stream which sweeps northward some miles offshore.

Westward the country rises and is entirely forested but never seems from the sea to be "mountainous." However, so far as the Connecticut River, it does have low mountains. The terrain near Thames River is sometimes quite precipitous. The coast is the southern terminus of a great glacier of eons ago and is extremely rocky from the inner coast northward. The base sub-terrain is subsiding slowly (18 inches in the past 1000 years).

The extended and narrow archipelago between Buzzard's Bay and Martha's Vineyard are the Elizabeth Islands, even today not very well settled. They have confusing currents and, being small, an unreliable water supply. They and Noman's Land Island are seldom visited. The original name of the island was apparently "Normans Land."

Pawcatuck River has little relevance to this study and is mentioned here merely to define the western edge of the current State of Rhode Island.

Thames River, just west, is significant because it was the site of a colonial settlement. Its governor was John Winthrop, son of a Governor of Massachusetts Colony. Records of the two places and individuals are sometimes confused by historians making it seem that the Massachusetts Colony surrounded Rhode Island. The confusion is compounded because of there being another Connecticut settlement along the Connecticut River, further west. This Thames River settlement had great bearing upon the history of the Narragansett Indian tribe.

More than 20 other serious studies have speculated that Leifsbudir was in New England and most others consider Karlseffni's Hop in New England or further south.

Artifacts throughout New England have been discovered since colonial times, distributed in such a way as to give plausibility to this hypothesis. At least two runestones are accessible and observable and others are recorded.

The site itself yields interesting information, but there is still much to be investigated and debated. This includes stone ruins still extant; traces of underground sites called "ghost fields" visible on particular satellite high altitude photographs of the type used by archaeologists to determine where old ruins exist; and historical records indicating various anomalies.

The locus of the Narragansett tribe was directly within and about the south-flowing Pettaquamscutt River. In 1986, archaeologists surveyed the river basin lowland areas. Researchers found that the valley had been essentially unoccupied for a very long period and reoccupied by a people of varied and advanced land use "about a thousand years ago." I will discuss this later but this startling information coincides with the Vinland voyages. This is where I believe the Vikings interacted with the Indian tribe that became known as the Narragansetts.

Little can be done without first coming to terms with what the sagas have to say. They say quite a bit; and for those who wish to take the trouble and who have other fields of expertise than seamanship, certain events and descriptions hold promise of even more insights. It is my belief that the essential opportunity to deal with the sagas should be presented to all.

I have given the sagas, long since in the public domain, here with my own interpretation in some cases. The sagas themselves are an Icelandic treasure, still preserved and treated as such. Additional sources reside in the folklore and private histories of Iceland and Greenland. It is my purpose to disseminate the tales in this mode and to indicate details available about possible sites used by the Vikings.

From these sagas, we can identify no fewer than 20 landfalls on the landmass of Vinland itself. My book aids immeasurably in perceiving the long sought land of Vinland. I believe that the prevalent view of Vikings as barbaric and murderous ruffians should incorporate that instead they were possibly the most advanced people of Europe of their day. Even their slaves had stipulated rights and the common person had recourse to courts of justice.

I have tried to present the seafaring problems of the travelers. I have also described how at least five vessels and more than 300 persons partook in voyages to Vinland. My main purpose is to bring this whole argument to debate. Only through mutual respect, intelligence, and open discussion can we contribute to the tale that may enthrall future generations as it has past generations for more than 400 years.

Yon high horizon, void appearing, mere nether barricade of mystery beyond!
Ours—if striving keeps us found.
Steadfast Leifur's chaffing courser waits at voyage end.
She—illustrious vessel that transited Europe and wracked on a spit in America;
She—that made Vinland thrice or more times;
We who follow in her wake—have only to traverse but once!
Gentle surge of morning tide—swirl and suck of wash 'mongst tidal rock can indeed
Become boom and crash of devastation,
But this ship of senses can never founder nor becalm.
Our literary vessel and grand endeavor may yet prove ranging—enduring as another.
So come aboard!
Find your place, fore or aft; test your gear and oar.
Resolve your mind to the trial at hand—the ship is bare afloat.
Her bow sways and bobs;
Stern shivers the binding sands those mites of movement connoting
A last, light thrust will set us on our way!
Feel the thundering, shuddering, streaming heartbeat of the world,
Of history beneath your feet. Look!
Multifarious shore-watchers count reflective few.
Cognitive ones who fathom our industry—await our cargo and our manifest.
Come aboard!
Our crew holds ready to row offshore; supplied sufficient; stowed; secure;
Pilot confident, guided, charted: sail lashed and set to hoist—
Billowing flaxen dray straining wild in billow to chase white horses of the Tumbling, running, rumbling sea.

Minds and souls fly and soar 'mong nebulous seabird throngs;
With dolphins and whales that sweep through our wash.
This Wave Cleaver was not built to slow for fog nor storm nor breaking wave.
Come aboard!
That spray from windward cannot afflict; the cold of it invigorates;
Its strike—e'en ice—but pleasurable sting;
The white of it a brightness in the night time—
The flight of it—the beat of angel's wing;
The taste of it the salt of life, the bulk of it bailed light with
'Boldened, gladdened hearts.
Who can say but it is spindrift flung from shaken locks of gods—of Neptune?
There! Look there!
Land rising 'neath towering clouds—through mists 'cross slow, slick swell of time?
That low and forest shore—a pleasant refuge—
Or island yielding rest, sweet water, new orientations?
Land? That cryptic legend land we seek?
There! There! Just o'er the bows!
That gleam of green; gray hovering clouds;
That arc of rainbow; glint of sunbeams bursting through?
Our lookout cries it, our pilot spies it, and crewmen yearn it!
Can it be our goal at last? Fair haven Vinland?
Leif's mystic Vinland—The Good?
Ahoy! Our port of call is yon!

"Capturing tempo and semblance of Sea and History, Fred N. Brown III"

The Coast of Vinland Explored from Space and Approached Anew

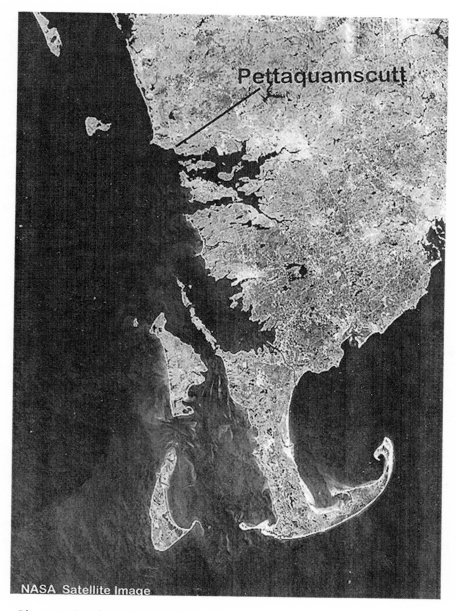

Photo 1: Southern New England—"Vinland"—from a coasting seaman's viewpoint. Oriented to fit the perspective of an explorer from the north—turning the corner, so to speak. (NASA photograph)

A seaman is always more interested in islands and shorelines than interior topography. Let us all be Vikings while we peruse this satellite photograph. The right hand border splits Boston Harbor in two. Along the coast to its south (left) can be seen the large estuary (Thorvald Erickson called it a fjord) of Plymouth which has a shallow light sand bottom, 14-foot tides and, I believe, is the "Crossannes" of the Vinland sagas. Listed below are the islands with sequential names from newest name to oldest.

Cape Cod (bottom) points north, as did a well remarked "north pointing cape" ("Promontorium Winlandiae") in the sagas. The area of the "elbow" is "Nauset." Extending south is the low sand spit island "Monomoy." There is evidence that this island did not exist a thousand years ago and there is easily enough white sand in the entire Cape Cod area to qualify as the "Wunderstrands" of the sagas.

South (left) is Nantucket which cannot be seen from Nauset area. Therefore, a ship exploring must turn west (up). The milky appearance of Nantucket Sound and environs shows the shoal conditions of it today. Up until 1840, Nantucket Harbor was a deep-water port. Thus the shallows, as well as Monomoy Island, are recent.

North of Nantucket (small islet is "Tuckernuck") is the large island "Martha's Vineyard," sometimes called "Martin's Vinyard" or "Capowock" or "Nope" or "Straumney," which is of a size to consume three days exploration as the sagas say. The far end of it is "Gay Head" and just south (left) of that is the small island "NoMansLand" (possibly "Norman's Land"). This place is of vital importance as it is an important consideration for a navigator and does, indeed, hold a runestone upon the southern shore.

The archipelago inshore paralleling toward Gay Head is called "Elizabeth Islands," three of which are named, "Naushon," "Pasque I.," and "Cuttyhunk." Extraordinarily strong tidal currents flow in the straits between them and Martha's Vineyard—so strong that the effect on the bottom sands can be seen from space. Other milder currents resulting from offshore Gulf Stream flow east along the southern shores of Martha's Vineyard, carrying enormous quantities of eroded sand into Nantucket Sound.

A distance further west (up) is the offshore island "Block" also called "Luiza" or "Sweetwater." To its northeast lies (right) the large estuary known as Narragansett Bay. This view shows how much a misnomer this is—surely a fjord (to a Norse seaman, at least.)

Further to the SSW is the tip of Long Island now referred to as the Montauk area and to its (right, near the shore) north is (partial view—five miles long)

"Fisher's Island," which I strongly suspect is the "Barn Island" landed by Thorvald Erickson in his exploration with the "afterboat" west of Leifsbudir. It is of a size and range (25 miles from Point Judith) to be ideal for the description and tone of the sagas.

To Block Island's NNE, the small peninsula pointing south is "Point Judith," a mariner's and meteorologist's datum. Along the coastline into, but not quite in, Narragansett "fjord" is an evident small indent into the land (a true but shallow "bay") and it is at the north end of this bay that Narrow River (outlet for Pettaquamscutt River) enters the sea.

The landsman's eye might be improved by knowledge that Nantucket cannot be seen (sea level) from the Nauset area, nor can Martha's Vineyard Straits be detected from near anywhere inside Nantucket Sound. From this, it is possible to infer the courses of any exploring seaman in this area. Indeed, it is possible and plausible to trace all of the detail and many of the (at least 18) remarked landfalls of Viking ship traffic along this coast a millennium ago.

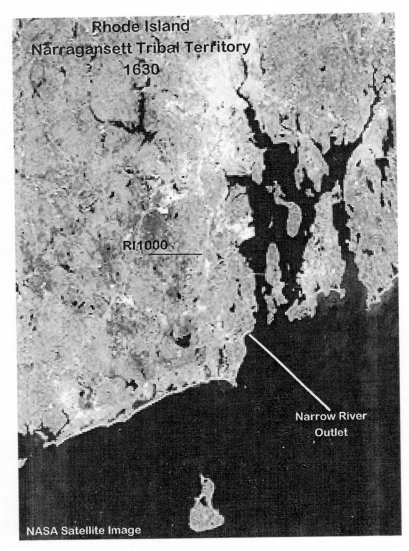

Photo 2: Rhode Island Narragansett Tribal Territory: Territory of Narragansett Indian Tribe near 1630. (NASA photograph)

At lower center is Block Island. My claim is that Leif Erickson approached this from the east, landed, and discovered a spring of clear and pure water. The saga reference is less than clear on this matter because of vagaries of narration, but the seaman will recognize the event from attention to water supply aboard ship. By coincidence, Block Island does have a number of springs and one was to become so famous for its purity in modern times that it became the site of a hotel and spa. Leif's departure was described as not quite north, and Narragansett Bay appears here as clearly NNE.

Block Island has all the feel of an offshore island and the horizon is broken at only one place along its 360 degrees. The small "spur" upon the coast is Point Judith and from this satellite photo it becomes obvious how natural is the course of an explorer from Block Island up into the bay to pass near Narrow River outlet. The only place any land at all can be seen from Block Island is a high point somewhat north of Point Judith named McSparran's (or Tower) Hill, somewhat more 400 feet high. This is a natural magnet for an explorer and is near the northern terminus of Pettaquamscutt River.

In 1524, Giovanni Verrazzano approached it (his "Luiza") from the west (New York Harbor). Since he had no maps, no pilot, no knowledge of the coast whatsoever, he had little choice but to make the same course as Leif had 524 years before.

Here again the landsman's eye must be aided by the fact that the so-called "East Passage" of Narragansett Bay into Newport Harbor cannot be detected even in the presence of numerous modern navigational beacons. This "East Passage," by way of which Newport is approached, is a misnomer of local familiarity. It is actually the central passage and the true East Passage is further east and is now known as Sakonnet River. Note the very narrow straits near the north end.

The two major islands adjacent within the "fjord" are (west) Jamestown (Conanicut) and (east) Rhode (Aquidneck) whereon lies the city of Newport. The large island north neatly fitting between them is Prudence, which gives us quite an improved vision of the stature of Roger Williams with the Indians. This sizeable island was also granted to him in addition to the huge areas of Providence and Pawtucket, which might be visible in this photograph as the light area at the tip of the fjord.

Just to the left of Point Judith is a large estuary called Great Salt Pond, which is among many misnomers in this area. It is a rather large tidal inlet, rich in fishing opportunities. It has always been an attraction to fishermen even today and has an advantage that Pettaquamscutt does not, for it supports prolific quantities of shellfish—a favorite food of Narragansett Indians. In southern Rhode Island, springtime life begins at this place. It is a less defined rift in the glacial moraine and its essence continues north in the form of marshy and broken land some ten miles or so. It has a portage over into the south end of Pettaquamscutt Lake and it was, in my opinion, along this portage that the saga "skraelings" traveled in that long ago springtime to the Viking settlement.

The place now has a considerable commercial fishery but very strong tidal currents in its narrow entry. Two opposing settlements guarding this entry are Jerusalem and Galilee. Power craft speed up or slow down around this area, but come and go at will, while sailing skippers must practice patience. This entry is now protected by a huge stone artificial breakwater, but even without this breakwater, the small entry is difficult, if not impossible, to detect by a ship's lookout entering Narragansett Bay. RI 1000 denotes a critical archaeological site of an Indian burial ground where tuberculosis lesions were discovered on a number of interments.

Photo 3: Two views of Narrow River. (Photos by Erick Brown)
Top: From inland toward the sea—almost the whole of it.
Lower: The rest of it from the same spot, a view toward a whole new continent.

Two views of Narrow River, which extends only about 5/8 of a mile, nearly a perfect kilometer. An industrious oarsman can walk it in nearly ten minutes or row it in five minutes.

The top view is outward toward the sea with the far dune being upon the sandbar. Despite the sometimes turbulent surf conditions just beyond that dune, this section of the river is normally placid and subject to disturbance only during extreme storms, and even then, a ship is relatively safe here. The woods to left on the north shore compare with saga descriptions of "forest coming near the shoreline."

The lower view is seen looking inward toward the lake and hill beyond. This is the view that almost certainly was seen by the small crew of the afterboat that preceded the ship left aground nearer the sea. Again, this is low tide in an unusual place, for this small river and lake has a tide of some 18 inches—somewhat delayed beyond regular lunar tides. The hill in background does not look as high and steep as it really is, although it lowers to the left (south) and rises northerly to Tower or McSparran Hill to the right. On the near right side of this view, archeologists have discovered traces of several ancient temporary Indian encampments used, apparently, by transient fishermen. Despite the small size of this river, it has an amazingly active submarine traffic. Divers have speared fish three feet long here. Runs of salmon and halibut are seen, with the salmon going to the north extreme and halibut only into the lake to spawn. A type of herring called "buckeyes" or "alewives" appear at the far end of the complex (certainly passing through yet never seen in this section of the river) where they force their way up steep rivulets into a lake some 50 feet higher than Pettaquamscutt.

In its pristine state, which was the condition when Leif Erickson saw it, it was forested with a variety of immensely big and high trees. It was likely clearer at the forest floor then, but nowadays is choked with underbrush. The area is profuse with both wild and feral grape vines, which sometimes blanket the underbrush and often climb to great heights into the treetops.

The greatest change is in the wildlife. In times past and partially into the life span of this writer, there were deer, rabbit, bear, wildcats of several species, puma (or mountain lion), snakes of many types including rattlesnakes, wolves, turtles up to three feet across, turkey, transient bird migrations, fox, skunk, etc.

When Leif arrived, it was vacant of aborigines, and this compares with archaeological evidence. Yet, when Giovanni Verrazzano sailed past in 1524, it hosted a large and prosperous population, distinct in many ways from other natives nearby. This book explains the anomaly.

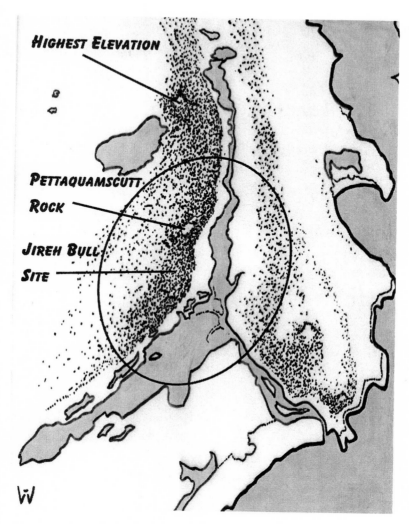

Photo 4: Jireh Bull Site and Pettaquamscutt Rock (Map by LuAnne Waddell with her brand marker of W)

Overall topographic conditions of the Pettaquamscutt River Valley and Jireh Bull's site. See discussion in Part Eight.

Photo 5: Oil painting of Ninigret, a chieftain of the Narragansetts (Used with permission of Museum of Art, Rhode Island School of Design, Gift of Mr. Robert Winthrop, Photography by Del Bogart)

Is this man a descendant of the Vikings? Among the many miracles that have come my way in this research is this painting of a Native American dating from that era when the two alien populations were in wary, but mainly amicable, contact prior to 1676.

There are a number of paintings of Indians extant, especially in Europe. Pocahontas, for instance, was portrayed formally in London, as were a number of other Indians who had traveled to Europe. However, this painting is unique in that it yields insight into the person, culture, and even genetics of a man who lived sometime in the period between 1620 and 1676. It depicts a place where relations between pioneer colonists and the aboriginal natives were friendly, and even welcoming on the part of the Indians.

Ninigret was a famed chieftain of a sub-tribe of the Narragansetts usually identified as a "Niantic." His home was but a mere 10 or 12 miles west of Pettaquamscutt and near the sea. There seems some uncertainty as to whether this was an older Ninigret in close relations with the Rhode Island and Connecticut settlers or a son who succeeded him. However, the context of history and ownership of the painting indicates that it is the famed Ninigret, who enjoyed an honored place in Rhode Island's history.

His home, a fort bearing his name, and his burial place have all been more or less preserved with all the vicissitudes of time and chance. Even his exhumed skeleton has been examined. The fort, located near the sea in Charlestown, Rhode Island, can still be seen in outline, and seems to have been square in form (i.e. European, probably of Dutch construction) with four battlements at corners. Accessible from the sea, it seems to have been a trading post for Dutch furriers at New Amsterdam (New York). A sword, which seems to have been Flemish, was excavated from there as well as several small cannons. I described the exhumation of his skeleton elsewhere in this volume. In life, he stood six feet, two inches tall, by chance precisely the same as Leif Erickson whose own skeleton has been recovered in Greenland.

I have examined this picture with a view toward noting differences that could be consistent with a Viking or European descendant. The Rhode Island Museum said that this is so important because it is our *only* contemporary rendering of an Indian near the time of first contact.

The painting is tremendously informative and contributes much to the theme of this work. It shows a man in good health, good spirits, and in noble demeanor. His dress compares exactly with the descriptions of Native Americans in this area left by Giovanni Verrazzano and Roger Williams. The stylized cape that he wears seems to be of some fabric doubtless obtained in trade. Williams stated that this cape was not a utilitarian item of dress but cultural and can be compared with the stylized forms that both Vikings and Scots wore. Otherwise, his dress, even to the high moccasins, is similar to Indians elsewhere even so far west as Apache and Navajo Tribes in Arizona.

He bears in his right hand a hewn round stick as if this were some sort of symbol of authority. The four-sided object in his belt may be a calendar stick for tracking dates and important events or it may be the handle of a weapon such as a knife. Indian calendar sticks were known all across North America so the idea must have been American, but it is remarkably similar to the same sort of device used by Vikings, which they called "primstavs."

His red loincloth may have been from trade but it bears noting that the color red was of intense interest in the Indian/Viking trading at Pettaquamscutt near AD1000. The material looks like it might have been leather, in which case perhaps the trade item was dye. This may have been simply personal preference but might also be an acquired attribute from Viking symbolism of red meaning war and white meaning peace.

The large medallion at his neck and the several lower on the necklace appear so light in hue that they might be either silver or pewter. If silver, it might be native, but if pewter, it was from trade. It could also be copper which compares with the Narragansett extensive use of this metal.

The black and white headgear and necklace are Wampum (peague) made of shell fragments and used as money. The amount seems less than that worn by full chieftains and in later years when it had a rate of exchange with English money, it was in standardized widths and usually rather long lengths as long as a "fathom" (6 feet) and longer. It was finely crafted and its value was regulated among all the local tribes.

Now for genetics. First is his stature, which is not apparent in the painting, but is so from his exhumed skeleton. He was 6'2" tall and both Verrazzano and Williams remarked on this racial attribute as compared with both their own races. Stature is a result of both genetics and interbreeding of races and we know the Vikings to have been tall.

Coloration: The artist painted his skin with a light bronze. However, the painting also seems to be dirty with a background so dim it is difficult to detail. Williams said that the natives were colored from the sun and "annoyntings" but were born white. Verrazzano said that they were "light to olive." I think the shading on the front of his thighs might give an accurate concept. Even so, many other depictions of Indians elsewhere show them as much darker than this. Several contemporary Mohawk depictions show nearly Negroid coloration and their locale was only some 125 miles away. This painting suggests lighter skin which may have some Caucasian genes from the Vikings.

Hair: The unpracticed eye might slip over this trait, yet it seems to me the most significant. I have shown this painting to a number of Indian friends and acquaintances here in Arizona (where they increase, still live, and sometimes prosper) and all show considerable surprise at what they see. All agree that his hair is unique and never seen elsewhere, nor have I ever seen anything in old photographs but the straight, coarse brownish hair expected from Mongolian descent. This hair shows as jet black/raven black, rarely encountered, if at all, in pure blood Indian cultures. The color seems to be a result of intermarriage probably with Vikings. It is typical of modern Mexican descendants of mixed Conquistadors and Indians. The texture is obviously finer (more Caucasian) and with a loft that is unique. It matches Verrazzano's description that the hair was "flowing" as it is here. Being a noble, his hair seems to be well tended and clean. The left side is in a tight braid and this is likely a military attribute permitting him to draw bowstrings.

Eyes: For a man 6'2" his eyes are enormous. I do not mean to imply that this is from Norse genetics but does, in my opinion, draw an important inference for the truth of this book's theme. At least three sources in the Vinland sagas remark on the large size of "Skraeling" eyes, twice from men and once, with emphasis, from Gudrid Thorbjornsdottir. Therefore, this is an Indian attribute, but also a local one. It seems to tell us that these Native Americans are the same as those encountered by the Vinland Voyagers.

There is a bit more information we can extract from this painting in looking over the background. Immediately behind the figure are several large rocks that seem to have been shattered in some manner. This shows the original condition of the lands in Rhode Island, for this is the remnants of a glacial moraine. The glacier was said to have been a full mile high! It dredged rocks and boulders from great distances and deposited them all through this coastline.

Ninigret stands at a place that looks familiar to me, except that the rocks seem to have been removed. Thirty years ago, intrepid explorers stood on a small beach on the north side of Narrow River and shared a rather lengthy visual conversation with a very alert, very curious, and beautiful fox—demeanor commanding, raiment luxurious, golden, glorious. The fox was comfortably seated—enthroned—while we three stood in raptured attendance at the sight this spectacle of natural beauty—an irreplaceable jewel, rarely experienced. Reynard seemed to ask, "Who are you, and what would you here in my domain?" Our answer lay in our wondered and pacific awe. After a few well considered minutes, accepted, we were left to our own quest. We faced a vista quite similar to what is seen here.

At lower left are indications that the ground falls away to a sandy place, likely a beach. In the background is the vision of New England's forest primeval where trees form bowers over a glen-like forest floor. It was certainly not Fort Ninigret for I have also spent time there and this place has no similarity to that area. It well might have been painted as far away as Mystic, Connecticut, where Ninigret's friend, military ally, and neighbor John Winthrop (the younger) was Governor. Centuries later, a donor who was a descendant of this original English settler donated this painting to the Rhode Island School of Design Museum.

The Coast of Vinland: Possible Locations of Leifsbudir, Hop, Straumney and Crossannes

A careful reading of the sagas suggests more than 20 possible landfalls upon the coast of Vinland that were potentially used by Vikings. I believe that the sagas were at least as true as Homer's *Iliad* and *Odyssey,* which were used to find Troy and other sites. Thorvald's detailed homeward voyage identifies a number of landfalls, of which the most important is Keelness, since this is a landfall in common with the following expedition of Karlseffni.

Thorvald observed that Crossannes seemed to them to be close to Leifsbudir. Most scholars who accept Vinland as being in New England identify Plymouth Harbor as Crossannes. Moreover, the closer one looks at it, the better it becomes. Plymouth Harbor has a place where the ship could be brought in against a steep bank for landing with a gangplank. This place is directly adjacent to a high and steep hill that compares with saga descriptions exactly.

The sagas suggest that all landfalls to the south of Crossannes must have been upon one landmass. The sagas state that Thorvald's motives were to explore Vinland. Coupled with the knowledge that Leif specified where and how it was to be done at his initial landing, Thorvald was most likely intent upon seaward explorations. This is why another trip westward took place by the Vikings in the even smaller after-boat.

Leif's saga listed seven landfalls: (1) wunderstrands, (2) stretch of broken coastline, (3) island of sweet tasting water, (4) land to south of it, (5) a fjord, (6) a river, and (7) Leifsbudir. He remarks rather specifically about a north-pointing cape as a bearing point that had been passed.

Thorvald's saga listed nine in addition: (1) an island west of Leifsbudir, (2) a rocky and wooded coastline to it, (3) a strand where the slaying of five took place, (4) Dungeness, (5) bird nest island, (6) Keelness, (7) a cape which did not have a specified size or orientation, (8) Crossannes, and (9) a river oriented east/west.

Since the latter river cannot be demonstrated as being upon Vinland, we will drop it from considerations. However, if the orientation is correct, it supports a trip the Vikings made upon one of the west coasts of Nova Scotia or Newfoundland.

Thorfinn adds six possible landfalls: (1) Wunderstrands, (2) a major cape, (3) Keelness, (4) Straumney, (5) Straumfjord, and (6) Hop.

This represents 22 landfalls but some may have been the same ones observed by others. Therefore, I can formulate a working hypothesis of what Vinland must have been like.

The Wunderstrands appear to have been repeated so we will reduce two later references to 19 landfalls. The north-pointing cape seems explicit in Leif's saga but less so in the others. For the moment, we cannot remove the possibility of more than one cape. However, indicators seem to show that the less definite descriptions suggest that the cape was pointing north.

Returning to the sagas, Karlseffni was intent upon locating Leifsbudir, which already had at least one large structure in place. He had negotiated for it as well. If the fjord that contained Leifsbudir were as close as Straumney, it would seem likely that they would go there promptly. Some saga descriptions say that it was "... a long way (or long time)" from Straumney to Hop. On the other hand, the trip back to the island after the battle (or the isolated reference that Karlseffni and 40 men went in advance) implies that it could not have been very far away.

As we said before, we are left with 19 landfalls to account for. These must be near extensive white sandy beaches and an island of a certain size and description. The size limits what we may consider, because there are not that many existing islands along the Atlantic coastline. The coastline was described as being oriented east and west, and Thorvald's expedition west supports that. It must have had a southerly exposure as Leif's stipulated course to Leifsbudir was to the north, which had a fjord.

In general, from the sea, the entire coastline of our projected "Vinland" of the original sighting seems "low and wooded" as was described by Bjarne Herjolfson near 986 A.D., but it takes on more complexities as one approaches closer. Outer (north jutting) Cape Cod is a low-lying sandy feature with low forest covering it. It is a geographic feature resulting from sea currents, mainly the north flowing Gulf Stream which sweeps some miles offshore. It was low and wooded in some sections but also broken and rocky (but still wooded) in others. Descriptions of the coastline and specifically Leifsbudir hold that the forests came very near the shoreline.

Some sizeable islands are Prince Edward Island, Mount Desert Island, Nantucket, and Martha's Vineyard. Block Island will not do because by sailing the distance, I found it to be only a one-day expedition. Prince Edward Island and Mount Desert Island do not have the correct bearing to match the sagas. Therefore, we might try to decide between Martha's Vineyard and Nantucket. Martha's Vineyard is an ideal candidate for Straumney. It fits perfectly in all respects—even to the strong currents that are indeed virulent enough to stop and even reverse a sailing vessel.

Therefore, as a pretty well supported theory, we can accept Martha's Vineyard as a base point to analyze other landings. The main landing, of course, was Hop, but perhaps that and Leifsbudir and Crossannes might be discovered from these descriptions.

Scholars are referred here to the early and comprehensive work of the Dane, Carl C. Rafn, who did the same analysis in 1837. Unless Karlseffni had inadvertently passed Leifsbudir, for which he was looking hard, both Hop and Leifsbudir must be somewhere west of Martha's Vineyard. Rafn also drew the same conclusion, placing Hop and Leifsbudir adjacent and within sight in the same fjord.

At this point, I postulate that Leifsbudir was in or near a fjord that had an island to its south or southwest. While Hop did not have such bearings to follow, the eventual discovery of the north/south river fits descriptions of both Hop and Leifsbudir! Therefore, I believe that Leifsbudir and Hop are the same place, only the residence of Leif and Thorfinn being apart by only a little distance.

Other Vinland scholars have come to this same conclusion. Leif never named his home site. We call it Leifsbudir, but he did not. There is no good reason why Karlseffni could not refer to it in any manner he pleased. In fact, it is not certain that Hop was an official name. It is descriptive of certain conditions that are translated as, "a lake into which salt water flows at high tide."

Karlseffni and Gudrid negotiated with Leif for residence at Leifsbudir. They set off in search of it. Their landfall at Keelness demonstrates that they were at least upon the coast of Vinland. Conditions at Straumney add to the supposition. The argument with Thorhall the Hunter confirms that they were still in search of it. It was unlikely that Leifsbudir could have been north of the cape, since it would have been missed twice in that case, where it had been easily found.

However, the most insightful matter that aids the case is Freydis, Leif's half sister. She was clearly a member of two expeditions to Vinland, once with the party of Karlseffni and later with her own expedition. She was a resident at both Hop and Leifsbudir—this much is explicit in the sagas. She was able on her own trip to go directly with two ships not sailing in company to Leifsbudir. If she had never been there before, how could this be possible?

Those interested in further examinations of southern New England might obtain several nautical charts from the U.S. Department of Commerce: "Nantucket Sound and Approaches," and "Martha's Vineyard to Block Island." The only thing lacking is exploring for Thorvald's island to the west.

A chart showing Plymouth Harbor, north of Cape Cod, would be helpful, for I believe that this shallow harbor is certainly the Crossannes of the sagas. For a seaman to travel east and around the cape, then west to Plymouth and not suppose he had returned near his origin would be strange indeed. Rafn agrees here.

Another interesting reference is an ancient map drawn by Cyprian Southack, published as a navigational chart in 1717, which is available from U.S. Library of Congress. The original lengthy name of Southack's collection of eight maps is:

> *The New England Coasting Pilot from Sandy Point of New York unto Cape Canso in Nova Scotia, and Part of Island Breton. With the Courses and Distances from Place to Place, and Towns on the Sea-Board; Harbours, Bays, Islands, Roads, Rocks, Sands: The Setting and Flowing of Tides and Currents; with several other Directions of great Advantage to this Port of Navigation in North America, by Capt. Cyprian Southack, who has been Cruising in the Service of the Crown of Great Britain Twenty-Two Years.*

Although greatly criticized, this map provided valuable information to treasure hunter Barry Clifford who was able to find the 1716 wreck of the *Whydah* in 1984 and some 100,000 pieces including the ship's bell.

Southack's map shows extreme changes that have taken place since he drew it. This coastline is in a state of subsidence, which can now be estimated to have dropped some 18 inches in a thousand years to an apparent sea level rise. Only 150 miles north in Maine, the land has risen some few meters in the same era, as it has in Norway. This curious anomaly is responsible for a major breakthrough in research of the site under study. Using Southack's map, I can now propose what the coastline of Vinland might have been like.

Cape Cod has long been noted as a hazardous area for shipping north and south, especially prior to construction of the Cape Cod Canal in the 1930s. The navigator must factor wind, weather, and currents at all times.

Southack's map shows no Monomoy Island at all, simply shallow soundings and the word "dry" at its terminals and varied places within Nantucket Sound, and surprisingly to the south and east of Nantucket. Accordingly, we presume that 700 years prior there must have been islets or islands, possibly with vegetation, at those places. Depths at the south/east shallows are now some 50/60 feet, probably because of proximity to the continental shelf. Otherwise, Nantucket Sound today is very shallow and has to be navigated carefully by way of buoyed channels.

A minor branch of the Gulf Steam, or an "influenced" current, flows parallel to the coastline and traverses between Martha's Vineyard (actually the smaller satellite Chappaquiddick) and Nantucket. This current, rather gentle, has some peculiar and dynamic effects. First, it draws sand to the eastward, sand that results from the extensive erosion of the outer coastline from greater or lesser sandy, high bluffs. Both Block Island and Martha's Vineyard have as their southern exposures relatively high bluffs of this type; Block is some 125 feet high, the Vineyard is a bit lower. It is apparent that this eroded sand created Monomoy Island and the shallows north of Nantucket.

Until 1850, Nantucket Harbor was "deep water," deep enough for larger whaling vessels, but now is restricted to small craft only. Therefore, the eastern end of Nantucket Sound is an area of deposition as opposed to the extensive erosion of inner and outer banks to the West. The deposition, by the way, is entirely of white (very light tan) sand. The entire makeup of the Cape Cod region is all this white sand—its most distinctive feature—and noted as comparable with the saga's descriptions of "Wunderstrandir/Furdirstrandir."

Until the construction of the Cape Cod Canal, steamships and smaller vessels such as schooners, barks, and barkentines traveled along and reciprocal of the inner currents and inside Long Island Sound. Larger and less maneuverable "square riggers" had to keep well clear of the entire area.

Lesser and more localized currents of interest are identified off Narragansett Bay, whose average four-foot tides invade and evacuate the huge estuary twice daily. They add or subtract some 20 percent of time to passages between Block Island and anywhere inside.

The currents shown in Martha's Vineyard Straits are most remarkable and of vital interest to us. Because of differentials in depths of deep Buzzard's Bay and shallow Nantucket Sound, tidal currents are set up in the area that are most unusual. They are extremely strong, sometimes reaching ten knots (my estimate) and are a genuine risk and problem to small craft. Sailboats cannot "buck" them and even powerboats must consider them seriously when traveling the area. I have seen them actually lay over and nearly submerge huge bell buoys designed to stay upright. Ferries of some 125 feet long have caused them to cut power and ride the currents for periods.

As a sailor, I can attest that appearances of what seem dominant features at sea are not as apparent as it would seem. For example, looking westward from Nantucket Sound, one cannot detect the presence of Martha's Vineyard Straits. It appears to be all land in that direction. Conversely, one cannot detect them from the west either, or from any point. One must pass the difficult elongated archipelago of the Elizabeth Islands and even then, the passage is not readily apparent.

The definitive work of Carl Christian Rafn included a map of this very area. His great work seems little addressed in this modern era, yet in its day was a very well known and influential study. Rafn, along with Finn Magnusen, his assistant, wrote and published *Antiquitates Americanae* in Latin in 1837.

The beginning of interest in Viking explorations in America dates from his publication by the Royal Danish Society of Northern Antiquaries. This large volume of old Icelandic documents, in which the proofs were set forth that the discovery credited to Columbus was anticipated by sea-roving Norsemen five hundred years earlier, was edited by Prof. C. C. Rafn. He was the founder of the Royal Danish Society, and a very distinguished antiquarian. Although Rafn never visited the United States, he received numerous drawings (it was too early for photographs) and information from various scientific entities in America.

His book appears to have been on the West Point Military academy curriculum near that time. I say that because Colonel (Brevet General) George Armstrong Custer in his well-written work *My Life on the Plains*, first serialized in the magazine *Galaxy*, quoted it.

Rafn's work was the culmination of collaboration between several New England Historical Societies and the Scandinavian principals in a serious attempt to explain the numerous artifacts and runestones known from colonial times, many of which could still be seen when he studied the area. There are eight runestones, identified as "inscription rocks" described on his map and in his book.

The wonderful Web site of the Northvegr Foundation comes to the aid of researchers. They have presented a summary of Rafn's immense book, which they allow to be freely distributed, and is placed here for clearer edification. A few excerpts from Rafn's history of the Vikings from sagas and actual documents are included here for clarification.

> Gardar the Dane, of Swedish origin, was the Northman who discovered Iceland in the year 863. A few out-places of the country had been visited previously, about seventy years before, by some Irish hermits ... Here, on this distant island-rock, the Old Norse language was preserved unchanged for centuries, and here in the Eddas were treasured those folk songs and folk myths, and in the sagas those historical tales and legends, which the first settlers had brought with them from their Scandinavian mother-lands. Iceland was, therefore, the cradle of a historical literature of immense value.
>
> The situation of the island and the relationship of the colony to foreign countries in its earlier period compelled its inhabitants to exercise and develop their hereditary maritime skill and thirst for new discoveries across the great ocean. As early as the year 877, Gunnbjörn saw, for the first time, the mountainous coast of Greenland. But this land was first visited by Erick the Red, in 983, who, three years afterwards, in 986, by means of Icelandic emigrants, established the first colony on its southwestern shore, where afterward, in 1124, the Bishop's See of Gardar was founded, which subsisted for upwards of 300 years ...
>
> On a voyage from Iceland to Greenland in this same year (986), Bjarne, the son of the latter, was driven far out to sea toward the southwest, and, for the first time, beheld the coasts of the American lands, afterwards visited and named by his countrymen. In order to examine these countries more narrowly, Leif the Fortunate, son of Erick the Red, undertook a voyage of discovery thither in the year 1000.
>
> He landed on the shores described by Bjarne, detailed the character of these lands more exactly, and gave them names according to their appearance: Helluland [Newfoundland] was so called from its flat stones; Markland [Nova Scotia] from its woods, and Vinland [New England] from its vines.
>
> Here he remained for some time and constructed large houses, called after him Leifsbudir (Leif's Booths). A German named Tyrker, who accompanied Leif on the voyage, was the man who found the wild vines, which he recognized from having seen them in his native land, and Leif gave the country its name from this circumstance.

I would like the reader to have the exact passage describing this find excerpted here from the *Flatey Book*. The same passage also includes the extraordinary observation about the length of the day, which places a particular latitude on the site:

> There were so good land qualities that it seemed to them that there might be no cattle fodder wanted in wintertime. There came no frost in winter, and little with-

ered the grass. More was there equal length of day and night than in Greenland or Island. The sun had there Eyktarstad (sunset after 3 p.m.) and Dagmala Stad (sunrise before 9 a.m.) on the shortest day ... Then Leif said to him: why wert thou so late, foster-father mine [Tyrker], and partedst from thy comrades. He spoke then first a long time in his southern tongue, rolled much his eyes, and made wry faces, but they did not understand what he said. He spoke then in northern language after a while: 'I have walked not much farther, yet I can something curious relate: I found wine trees and wine-berries.'

Now I will return to the Vegr Foundation's description of Rafn's book:

Two years afterward, Leif's brother, Thorvald, repaired thither, and in 1003 caused an expedition to be undertaken to the south, along the shore, but he was killed in the summer of 1004 on a voyage northward, in a skirmish with the natives.

The most distinguished, however, of all the first American discoverers is Thorfinn Karlseffni, [Karlsefne] an Icelander, whose genealogy is carried back in the Old Northern annals to Danish, Swedish, Norwegian, Scottish, and Irish ancestors, some of them of royal blood. In 1006 this chieftain, on a merchant voyage, visited Greenland, and there married Gudrid, the widow of Thorstein [son of Erick the Red,] who had died the year before in an unsuccessful expedition to Vinland.

Accompanied by his wife [widow of Thorstein Erickson,] who encouraged him to this voyage, and by a crew of 160 men on board three vessels, he repaired in the spring of 1007 to Vinland, where he remained for three years, and had many communications with the aborigines. Here his wife Gudrid bore him a son, Snorri, who became the founder of an illustrious family in Iceland, which gave that island several of its first bishops. His daughter's son was the celebrated Bishop Thorlak Runólfson, who published the first Christian Code of Iceland. In 1121, Bishop Erik sailed to Vinland from Greenland, doubtless for strengthening his countrymen in their Christian faith.

The notices given by the old Icelandic voyage chroniclers respecting the climate, the soils, and the productions of this new country, are very characteristic. Nay, we have even a statement of this kind as old as the eleventh century, from a writer (not a Northman), Adam of Bremen. He states, on the authority of Sven Estriðson, King of Denmark, and a nephew of Canute the Great, that the country got its name from the vines growing wild there. It is a remarkable coincidence in this respect, that its English re-discoverers, for the same reason, named the large island, which is close off the coast as Martha's Vineyard. Spontaneously growing wheat (maize or Indian corn) was also found in this country.

In the meantime, it is the total result of the nautical, geographical, and astronomical evidences in the original documents, which places the situation of the countries discovered beyond all doubt. The number of days' sail between the several newly found lands, the striking description of the coasts, especially the white *sand-banks of Nova Scotia and the long beaches and downs of a peculiar appearance on Cape Cod (the Kjalarnes and Furðustrandir of the Northmen), are not to be mistaken.*

In addition hereto, we have the astronomical remark that the shortest day in Vinland was nine hours long, which fixes the latitude of 41 deg. 24 min. 10 sec., or just that of the promontories which limit the entrance to Mount Hope Bay, where Leif's booths were built, and in the district around which the old Northmen had their head establishment, which was named by them Hóp.

When we reflect that the strongest and most undeniable evidence has been adduced to prove the fact of an Icelandic-Norwegian discovery of our continent in the tenth century, it becomes a matter both of surprise and regret that some of our most lauded writers of American history should either leave the event entirely unnoticed, or have disposed of it in some half-dozen well-rounded and skeptical sentences.

Quite scholarly, *Antiquitates Americanae* should be recommended reading for serious Vinland scholars. Otto Zeller republished it in Germany in 1968 with commentary in German. Zeller's book added more information about Dighton Rock, which is the causative agent for some to believe that Hop was at Bristol, Rhode Island.

I think very highly of Rafn, and what were considered ridiculous theories of his were confirmed by the finding of the Viking site at L'Anse aux Meadows. I do, however, have some objections to his precise locations.

Rafn places Leifsbudir in the Tiverton area of Massachusetts and Hop in the Mount Hope areas near Bristol, Rhode Island. His entry to both these places is up the so-called Sakonnet River, which, however, introduces a first doubt of the theme. For the waterway, truly the correct eastern passage of Narragansett Bay is in no way a river. It is salt and has no fresh water current anywhere in it. This is at variance with the saga description.

His descriptions of runestones and the Newport Tower have been criticized. He can probably be criticized because he took the word and drawings of others and did not view them himself. Would he have thought them all genuine if he had come here to examine them?

An important item to address is his location of "Krossannes" (which I spell "Crossannes") here shown as near the Plymouth Harbor entry. An amazing consensus of Vinland scholars agree on this locale and it would seem the most clearly defined of Vinland sites.

One scholar who claimed to believe Rafn's hypotheses was William Goodwin. A cautionary note is offered about William Goodwin and his book, *The Truth about Leif Ericsson and the Greenland Voyages to New England,* Meador Press, Boston, 1941. He published a map, which resembles Rafn's and at first glance seems identical—in fact, he attributed the map to Rafn. In Goodwin's map, Krossannes is moved up to Boston Harbor to bolster his theories because he bought Mystery Hill in 1936 in Salem, New Hampshire. He developed that site for tourists. He called it America's Stonehenge but there is no record of anything connected with the Vikings. My hope is that Mr. Goodwin's research source was responsible for the grievous error.

The site I visited and believe to be Vinland resolves the question. To the west of Martha's Vineyard or Straumney lies the smaller island of Block. To its north/northeast lies the fjord we call Narragansett Bay. Scholars including even Dr. Helge Ingstad have examined this bay for the presence of a river that might fit the saga descriptions. Rafn theorized certain conditions on the east side of the bay/fjord. Dr. Ingstad wrote in his first volume that he could not find the river and so directed his attention northward to his great discovery in Newfoundland.

However, it was my good fortune to have found a sterling candidate river. It is difficult to find because of its unusual location. It does not lie within Narragansett Bay/Fjord where many have looked, but along the approaches on the west side. Nevertheless, the great explorer, Giovanni da Verrazzano, found it.

Small, almost insignificant, the landing by sea is actually the first practical one along a course entering the great waterway to the north. It is ideal for landing by a ship with oars and a large crew. It matches conditions described by both Leif Erickson and Thorfinn Karlseffni; i.e., that Leif's landing was in an attractive place and that Karlseffni's Hop river could not be entered except at high tide.

It is the Narrow River/Pettaquamscutt River, which lies at 41deg 25min north and 71deg 26min west, some 2,100 miles from Greenland and the same latitude as Oporto, Portugal; Rome, Italy; and Istanbul, Turkey. It also immediately resolves that puzzling observation that the river of Hop "... flowed down the land into a lake, and then into the sea." The complex does just that, with two distinct rivers with different names. These are Pettaquamscutt River and Narrow River, the first being some 7 miles north/south and the latter being less than a mile or a ten-minute row by a small boat oriented northwest/southeast. A topographical map of the area is available from the U.S. Geological Survey: Narragansett Pier, RI, N4122.5-W7122.5/7.5.

It is accessible and pleasant. Fishing is popular. The catch is likely to be salmon or even small halibut, just as in the Vinland sagas. Visitors should be advised that some foot traffic is permissible at the lower levels near the river. These areas are preserved for wildlife and owned by varied wildlife organizations. All the upper levels away from the rivers are private property and owners are sensitive to intrusions on their property. If you go, please respect the subject of this study and the rights of property owners. Do no harm. Leave nothing but your footprints; take nothing but your time.

One might be curious as to whether any artifacts or other indicators have been found by this historian or others in the past. Yes, the site is historical because of certain colonial events there. A siege took place at the old Jireh Bull property and garrison, which is now owned by the Rhode Island Historical Preservation Society. It is on an elevation overlooking the lake. It was archaeologically surveyed in 1917 and two structures older than 1663 were discovered before World War I interrupted the survey.

Aerial satellite photographs in my possession have been analyzed by the University of Texas with discovery of at least one "ghost field" or trace of three structures of some kind. On site, nothing can be seen above ground. Underground, there may be at least two other structures, but I am in no position to investigate the area where these undetermined structures were located. I will discuss them more at a later point.

The cliffs to the north are Pettaquamscutt Rock—a most impressive or even majestic outcropping just about a mile from the settlement above the lake. It seems ideal to me; 60 feet on the east side, 15 feet on the west; sheer all around and easily defended. Its flat top is about 150 feet square.

In 1889, an artifact was accidentally discovered some five miles away near the salt-water bay. It was turned over to a local educator who soon became convinced that it was a Viking battle-ax. James Earl Clausen was a high school teacher who speculated that the river had been the Hop of the sagas, and his discovery made the newspapers of the day. He described the ax as weighing ten pounds, and was eleven inches along the sharp edge and eleven inches from edge to haft hole. He was ridiculed and harassed but eventually wrote a manuscript about it. However, the manuscript was lost before publication and the ax seems to be lost as well as I cannot locate what happened to it.

The Coastline as Developed by the Voyage of Wave Cleaver

On the voyage of *Wave Cleaver*, I identified and studied the Pettaquamscutt River/Narrow River complex. In truth, this is a rather spectacular glacial rift. Because of its heavy, but low, forestation, its steep walls are not readily visible. It is not frequently traveled and some local people have never even been within the valley. It is about seven miles long.

According to Roger Williams who came to the area in the early 1600s, Narragansett was a small and highly revered island by the Amerinds within one of two waterways, the western one being known as Point Judith Pond. That Pond is a highly developed port. Pettaquamscutt has been neglected because Narrow River has a very restricted entry. It can only be entered by a ship (sail or power) at high, slack water tide. This is what the sagas imply for Leif's landing. The Viking ship grounded out as the sagas say when they entered Hop.

This coastline differs from the coastline depicted in Rafn's work in two respects. I have relocated Leifsbudir and Hop to Pettaquamscutt on the opposite side of Narragansett Bay, and I believe that Keelness is on a minor peninsula instead of the outer Cape entire.

Let me clarify a few things. The name Narragansett is somewhat controversial, there being no village or town of that name. Roger Williams in 1643 identified it meticulously as a small island within one of the two estuaries, one being Pettaquamscutt and the other Point Judith Pond. I have found that according to Williams' description, Narragansett is truly the one at Pettaquamscutt—that small islet now known as "Gooseberry."

Crossannes is a very powerful identifier since we have the word from the deceased Thorvald Erickson in the saga. That enables analysts to identify the estuary and locate the course of the ship to its landing site where "they put out the gangplank and went ashore." The saga describes a very steep shoreline and events thereafter describe a high and steep hill climbed by the adventurous party. This landing site can be located within yards in the site I have surveyed.

Now we come to tracing the courses of Leif Erickson. The prominent North Pointing Cape so well established in most Vinland studies must be, of course, Cape Cod. Then, as now, it is such a dominant feature of coastal traffic that it always has been an overbearing consideration of the coastline.

Leif's courses are generally presumed to be reciprocals in the opposite direction of that of Bjarne Herjolfson's some 14 years earlier. These passages were said to consist of two-day, three-day, and four-day blue-water passages. This coincides nicely with those openings of the coast from Cape Cod to Nova Scotia, from Nova Scotia to Newfoundland, and from Newfoundland to Greenland. It should not be surprising that the ship might again find itself at this distance, for it must be recalled that this is the same ship and most probably many of the same crew who had been here before.

Nantucket Island is not visible from anywhere on the cape, being of rather low-lying terrain. Therefore Leif's courses must have been, from here, west and of a "coastal" nature. The view west does not invite closer examination. It appears as a receding coastline from inside Nantucket Sound.

Coastal sailing is the most challenging of activities for the blue-water seaman. Proximity to a shoreline, especially at night, is the most dangerous place to be. Usually, if the navigator is familiar with the coast, he stays far enough offshore to travel from headland to headland. However, if he is not familiar with the coast he will be encouraged to close in to discover finer features for identification. This is the role of an explorer, which, of course, Leif Erickson was.

All prudent skippers would insist on constant vigilance of the lookout at the masthead, probably having two there at all times and everyone else on board on the watch for hazards. It is not what you see that is dangerous here, but what you do not see—such as submerged rocks and shoals. Masthead lookouts have an advantage, often being able to detect channels not observable from deck level. What they might see in what is now Nantucket Sound, if it were bright sunlight, would be wide expanses of green water with the blue of channels extending southward around nearby Martha's Vineyard.

Normally, the navigator hesitates to travel between an island and a mainland or larger island. The space between is often "shoal." It's not called a "sound" for nothing, and prudent crews travel outside of any island.

The distance between outer Cape Cod and Narragansett Bay is about 75 miles and could be sailed in a day if desired. At night, the ship must travel well off shore and wait or land at some convenient spot. A landing might have been made on an island along here somewhere, possibly "Noman's Land Island."

Obviously, a thousand years ago, these islands were essentially uninhabited because they were hard to access, had limited resources, and the population was sparse. Vikings, with their ships and warlike nature, were possibly the first culture to avail themselves of islands for residence and waystops. Since landings on mainlands always entail risks of navigational and possible overwhelming and hostile populations, islands can be safer landing areas. That is probably why Leif landed on the island I call "Island of Sweet Tasting Water," since they remarked on the purity of a spring found low in the grass. The island must have been large enough for a sizeable watershed to support the spring. Block Island is large enough and even today supports a number of pure water springs.

At the Block Island landing, Leif would be well aware of a fiord on the mainland to his north, having passed it en route. The sagas stipulate that his course from this island was "... north to a fiord, across that sound that lay between the island and that cape that jutted northward," i.e. a course somewhat off north either west or east, and Cape Cod can be seen on a map and presumed to remain in an explorer's memory to the eastward.

As it happens, no land is in sight from Block Island except the heights behind Point Judith, which happen to be those rising behind Pettaquamscutt as well.

Pursuing this course, which is a natural one, the coastline appears featureless on approach except for a wide and open passage to the east of Point Judith. Indeed, even upon entry to this passage (now locally called the "West Passage"), one cannot see any land at all to the north—it appears as a broad opening.

However, as it appears here, there was no need for Leif to investigate this, for a few miles along that passage would appear a strand. At the north end of that strand, a small river now named "Narrow River" can be perceived. It is an attractive spot from both land and sea. For a seaman looking for a happy haven, it cannot be beaten for a potential safe anchorage so near to open surf.

Although pleasant in appearance, it is an awkward approach. The channel is so close to shore and so serpentine that it invites groundings, which the sagas say was the experience of Leif Erickson. The grounding would not be necessarily dangerous as the bottom is sand and, indeed, perhaps the grounding was deliberate.

Once ashore on that remarkable sandbar, in a mere ten minutes walk, that placid lake lies just below the steep hill of Pettaquamscutt. So sheltered is this lake by the surrounding hills that even in severe storms and hurricanes, damaging winds and high waves never enter.

By combining what limited information we have from the sagas with what we know of seafaring principles, these courses of Leif's appear both possible and plausible, offering no difficulties whatsoever.

The map shows the angle of approach to Narragansett Bay from Block Island as sharper than it is. It is actually only 10 degrees off due north. I read the sagas as that sound between Block Island and somewhat back toward Cape Cod.

Ancient maps give insight to conditions of the area of Leifsbudir. In 1764, cartographer Charles Blaskowitz was assigned to survey Narragansett Bay for the British military to determine its potential as a naval base. Blaskowitz's map was completed in 1777 and was so detailed that even the farms and the names of the farmers were included. His details were important in a 1778 battle between Count Charles-Hector d'Estaing and Admiral Richard Howe during the American Revolution.

His map gives an impression of the terrain of Narragansett Bay on a map. The hills, however, are by no means as steep as his map implies but it does show the difficulties of the area. It also lays to rest the question of whether Narragansett Bay could be referred to as a fjord. Geologically, it really is a fjord and this map shows that it also looks like one.

A map of the general terrain of the entire State of Rhode Island usually shows it as nearly flat because there are few really high places, but flat it isn't. Pettaquamscutt is left off the French map as insignificant and the state map shows the terrain there as marshy, which is a long way from the fact. Pettaquamscutt Basin has high and often steep hills to over a hundred feet all along it.

Part IV

▼

Observations of Narragansett Amerinds by Early Visitors to America

We consider the idea that perhaps Vikings influenced native aborigines during their supposedly "brief" settlement, a miracle of locating two perfect resources—intelligent, and educated, precise men with no ax to grind.

Giovanni da Verrazzano, 1524

In 1524, a mere 32 years after Columbus' first voyage, the Italian Giovanni da Verrazzano, in command of a French vessel in commission of the King of France, came to North America to explore the coast. He struck land near the Carolinas, turned first to the south, and then reversed to the north for extensive and detailed observations all the way to Nova Scotia.

He described briefly the natives of Chesapeake Bay and what is now New York Harbor. Leaving that area, he coasted along Long Island and eventually left it behind for an observed offshore island [Block Island] which he named "Luiza" after the Queen of France. From Montauk Point, that island is the only land in sight.

He passed it on either the east or west, described it as forested (not now) and apparently densely populated because of many smoke columns from fires.

Here he must have come to an impasse. Again, no land is visible from there except for a small sighting directly north and in the distance—actually 18 miles. Since he wanted to make contact with the shore, this was the only course for him to take.

Many historians say that his landfall was in Newport Harbor, but as it happens, an approach to land toward the only sighting does not take into account that the shoreline as it develops seems featureless. What does develop is a wide expanse of water directly north, which is the now called West Passage of Narragansett Bay.

Verrazzano mentioned that native vessels of "exquisite artifice" crewed by 24 men met him and led him to a sheltered landing where an island existed, and that the native women were directed over to that island while the men came aboard.

This landing must have been what is known as "Dutch Harbor." Dutch Harbor has a nearby island, which might be the one mentioned in Verrazzano's journal. He noted that the natives would not allow their women to board the ship and sent them to a nearby island while the men interacted with Verrazzano's crew. A plausible subsequent move might have been northeast across to the mainland to another harbor now named Wickford. This small harbor is too constrained for that necessity of sending the women to it. Dutch Harbor also lacks any incentive to explore southward because the open ocean can be seen over the low peninsula. However, the terrain to the south of Wickford does invite investigation as well as having a unique waterfall ("flowed down from on high") at a convenient distance, described in Verrazzano's notes.

He stayed over two weeks calling it a "refuge" and described the place and its inhabitants more fully than any other place on his explorations. The following notes are translations from the *Written Record of the Voyage of 1524 of Giovanni da Verrazzano* as recorded in a letter to Francis I, King of France, July 8th, 1524. Giovanni signed it as Janus Verazanus.

> We reached another land 15 leagues from the island, where we found an excellent harbor; before entering it, we saw about boats full of people who came around the ship uttering various cries of wonderment. They did not come nearer than fifty paces but stopped to look at the structure of our ship, our persons, and our clothes; then all together, they raised a loud cry, which meant that they were joyful. We reassured them somewhat by imitating their gestures, and they came near enough for us to throw them a few little bells and mirrors and many trinkets, which they took and looked at, laughing, and then they confidently came on board ship.
>
> Among them were two kings, who were as beautiful of stature and build as I can possibly describe. The first was about 40 years old; the other a young man of 24, and they were dressed thus: the older man had on his naked body a stag skin, skillfully worked like damask with various embroideries; the head was bare, the hair tied back with various bands, and around the neck hung a wide chain decorated with many different-colored stones. The young man was dressed in almost the same way.
>
> These people are the most beautiful and have the most civil customs that we have found on this voyage. They are taller than we are; they are a bronze color, some tending more toward whiteness, others to a tawny color; the face is clean-cut; the hair is long and black, and they take great pains to decorate it; the eyes are black and alert, and their manner is sweet and gentle, very like the manner of the

ancients. I shall not speak to Your Majesty of the other parts of the body, since they have all the proportions belonging to any well-built man.

Their women are just as shapely and beautiful; very gracious, of attractive manner and pleasant appearance; their customs and behavior follow womanly custom as far as befits human nature; they go nude except for stag skin embroidered like the men's, and some wear rich lynx skins on their arms; their bare heads are decorated with various ornaments made of braids of their own hair which hang down over their breasts on either side. Some have other hair arrangements such as the women of Egypt and Syria wear and these women are older and have been joined in wedlock.

Both men and women have various trinkets hanging from their ears as the Orientals do; and we saw that they had many sheets of worked copper, which they prize more than gold. They do not value gold because of its color; they think it the most worthless of all, and rate blue and red above all other colors. The things we gave them that they prized the most were little bells, blue crystals, and other trinkets to put in the ear or around the neck. They did not appreciate cloth of silk and gold, nor even of any other kind, nor did they care to have them; the same was true for metals like steel and iron, for many times when we showed them some of our arms, they did not admire them, nor ask for them, but merely examined the workmanship. They did the same with mirrors; they would look at them quickly, and then refuse them, laughing.

They are very generous and give away all they have. We made great friends with them, and one day before we entered the harbor with the ship, when we were lying at anchor one league out to sea because of unfavorable weather, they came out to the ship with a great number of their boats; they had painted and decorated their faces with various colors, showing us that it was a sign of happiness. They brought us some of their food, and showed us by signs where we should anchor in the port for the ship's safety, and then accompanied us all the way until we dropped anchor.

We stayed there for 15 days, taking advantage of the place to refresh ourselves. Every day the people came to see us on the ship, bringing their womenfolk. They are very careful with them, for when they come aboard and stay a long time, they make the women wait in the boats; and however many entreaties we made or offers of various gifts, we could not persuade them to let the women come on board ship.

One of the two kings often came with the queen and many attendants for the pleasure of seeing us, and at first they always stopped on a piece of ground about two hundred paces away from us, and sent a boat to warn us of their arrival, saying they wanted to come and see the ship: they did this as a kind of precaution. And once they had a reply from us, they came immediately, and watched us for a while; but when they heard the irksome clamor of the crowd of sailors, they sent the queen and her maidens in a light little boat to wait on a small island about a quarter of a league from us. The king remained a long while, discussing by signs and gestures various fanciful notions, looking at all the ship's equipment, and asking especially about its uses; he imitated our manners, tasted our food, and then courteously took his leave of us.

Sometimes when our men stayed on a small island near the ship for two or three days for their various needs, as is the custom of sailors, he would come with seven or eight of his attendants, watch our operations, and often ask us if we wanted to stay there any length of time, offering us all his help. Then he would shoot his bow, run, and perform various games with his men to give us pleasure.

We frequently went five to six leagues into the interior, and found it as pleasant as I can possibly describe, and suitable for every kind of cultivation—grain, wine, or oil. For there the fields extend for 25 to 30 leagues; they are open and free of any obstacles or trees, and so fertile that any kind of seed would produce excellent crops. Then we entered the forests, [other translations say "clearings"] which could be penetrated even by a large army; the trees there are oaks, cypresses, and others unknown in our Europe. We found Lucullian apples, plums, and filberts, and many kinds of fruit different from ours.

There is an enormous number of animals—stags, deer, lynx, and other species; these people, like the others, capture them with snares and bows, which are their principal weapons. Their arrows are worked with great beauty, and they tip them not with iron but with emery, jasper, hard marble, and other sharp stones. They use the same kind of stone instead of iron for cutting trees, and make their little boats with a single log of wood, hollowed out with admirable skill; there is ample room in them for fourteen to fifteen men; they operate a short oar, broad at the end, with only the strength of their arms, and they go to sea without any danger, and as swiftly as they please.

When we went farther inland we saw their houses, which are circular in shape, about 14 to 15 paces across, made of bent saplings; they are arranged without any architectural pattern, and are covered with cleverly worked mats of straw which protect them from wind and rain. There is no doubt that if they had the skilled workmen that we have, they would erect great buildings, for the whole maritime coast is full of various blue rocks, crystals, and alabaster, and for such a purpose it has an abundance of ports and shelter for ships.

They move these houses from one place to another according to the richness of the site and the season. They need only carry the straw mats, and so they have new houses made in no time at all. In each house there lives a father with a very large family, for in some we saw 25 to 30 people.

They live on the same food as the other people—pulse (which they produce with more systematic cultivation than the other tribes, and when sowing they observe the influence of the moon, the rising of the Pleiades [M45 or the Seven Sisters or to the Vikings, Freyja and her hens—seven stars that shine brightest in June], *and many other customs derived from the ancients), and otherwise on game and fish.*

They live a long time, and rarely fall sick; if they are wounded, they cure themselves with fire [cauterization] without medicine; their end comes with old age. We consider them very compassionate and charitable toward their relatives, for they make great lamentations in times of adversity, recalling in their grief all their past happiness. At the end of their life, the relatives perform together the Sicilian lament, which is mingled with singing and lasts a long time. This is all that we could learn of them.

This country is situated on a parallel with Rome at 40 2/3s degrees, but is somewhat colder, by chance and not by nature, as I shall explain to Your Majesty at another point; I will now describe the position of the aforementioned port. The coast of this land runs from west to east. The harbor mouth [his footnote: "which we called 'refugio' because of its beauty"] faces south, and is half a league [half a league must mean west passage] *wide; from its entrance it extends for 12 leagues in a northeasterly direction, and then widens out to form a large bay of about 20 leagues in circumference.*

In this bay there are five small islands, very fertile and beautiful, full of tall spreading trees, and any large fleet could ride safely among them without fear of tempest or other dangers. Then, going southward to the entrance of the harbor, there are very pleasant hills on either side, with many streams of clear water flowing from the high land into the sea.[Another translation describes the water *"flowing down from on high"* as a waterfall might do.]

In the middle of this estuary there is a rock of "viva petra" [living rock] *formed by nature, which is suitable for building any kind of machine or bulwark for the defense of the harbor.* [His footnote: *"which we called La Petra Viva, because of both the nature of the stone and the family of a gentlewoman; on the right side of the harbor mouth there is a promontory which we call Jovius promontory."*]

Let us add something about this translation. The original note on Verrazzano's letter was "promontorio jovio" which we have learned means "majestic headland or promontory." "Promontorio" is translated into English as headland or high land and into Old Norse as "ness." Jovio and Jovius come from the name of the gigantic planet of Jupiter and therefore mean "majestic." Hence, this phrase means "majestic head land or high land." Could this have been where the Vikings fled when attacked by the Skraelings?

The other important designation is a map from Vesconte de Maggiolo (or Maiollo) in 1527. He has a map with the designation "Norman Villa" many miles above Luiza or Block Island, and six lines below that is the designation "refugio," from Verrazzano. In other words, Maiollo's 1527 map shows a "ville or villa" in the vicinity of New York Bay since it is so far away from the notation "refugio." Maiollo was not there and was using the report or map of presumably Verrazzano so it is possible that his distances were a bit off.

The notation "Norman Villa" on Maiollo's 1527 map has made many believe that either the Newport Tower or some other prominent structure(s) was noted. "Villa" can mean an estate or a country seat (village) smaller than a city. "Norman" derives from Northmen or Norsemen who came to that area of France now called Normandy but it can also refer to Norsemen who came to other areas such as England or Italy.

Girolamo (Giuseppe's brother) da Verrazzano produced a 1529 map that had no such designation as "Norman Villa" but south of Luisa (Block Island) had the comment "B. del refugio" at the exact point of Narragansett Bay, matching the same designation as "refugio" on the 1527 map of Maiollo.

The "viva petra" or living rock is probably Whale Rock, an exposed reef visible to the northeast from Leif's river mouth. It very much resembles a breaching whale and the waves breaking over and around it make it look very much alive. This reef is so dangerous that it had its own lighthouse, which happened to be nearly adjacent to the river mouth. This lighthouse, built of cast iron, was completely washed away in the great hurricane of 1938 and very little wreckage of it ever found, nor the body of the unfortunate keeper. This storm, the big event of my generation and place, was extraordinarily violent and the worst of a series that originated then and continues periodically to this day. I can make claim to have weathered no fewer than 12 and of these, five afloat.

The natives Verrazzano described were probably Narragansetts and not Wampanoags as many historians claim. Both tribes are, however, of the Algonquian culture. We are much indebted to this episode of history that opens up knowledge of the Narragansett Indians, whom Verrazzano stated were taller than themselves (French sailors) and of light coloring.

Estevan Gomez sailed within months of Verrazzano's visit to this same area after visiting Nova Scotia and Prince Edward Island. He went along the North American Coast and named the Penobscot River "Rio de las Gamas" because of the plentiful deer along its banks. He found the natives to be friendly and the country "temperate and well-forested," with oak, birch, olive, and wild grape, but on his carefully drawn map, he wrote, "No gold here." Realizing the Penobscot River was not the strait to China that he sought (as did all European voyagers of that era) but "a famous river with a great flow of water," Gomez continued along the North American coast. He explored the Rio de Juan Bautista (Boothbay), the Rio de Buena Madre (Kennebec River), the Rio de San Antonio (Merrimac River), and the site of present-day Newport, Rhode Island. It is possible he went as far south as New Jersey.

In 1525, unable to bring back gold or treasure, he filled his ship with Indians in the hope that his king, Charles V of Spain, would consider slaves to be valuable. Samuel Morison believed that because the Indians in Newport were so friendly to Verrazzano the year before, that might have been where he picked up his captives.

Gomez drew an outline map, which was embodied in the 1525 planisphere of Diego Ribero (geographer for Charles V). When again, five years after, the most distinguished geographers of Spain and Portugal met to settle disputes arising out of Pope Alexander's grants, the outlines of America were fixed for the first time from the discoveries of both nations.

In Ribero's 1529 chart, the country from Maryland, New Jersey, New York, and Rhode Island is called the "Land of Estevan Gomez." B. F. DeCosta described this information in an 1869 article "Who Discovered America?" in *The Galaxy Magazine*. DeCosta's source was the 1559 *Historia General de Las Indias* by Francisco Lopez de Gomara published in Barcelona. Georg Michael Asher also described some of this in his account of *Henry Hudson, the Navigator*.

Ribero's map also detailed areas explored by those voyagers mentioned earlier such as Luis Vasquez de Ayllon and Gaspar and Miguel CorteReal. Unfortunately, we have no description of Gomez's landfall (if any) in Rhode Island other than this scanty information.

Narragansett Tribal Lands When Roger Williams Arrived in 1635

As an interesting aside, Captain Christopher Jones who brought settlers to Plymouth on the *Mayflower* had sailed it to Norway in 1609 to trade timber for fish. He is also reported to have traded with Iceland as well, according to Nathaniel Philbrick and others. However, by the time Jones brought settlers on the *Mayflower* in 1620, a devastating epidemic struck the aborigine populations of the eastern seaboard in 1615.

Some theorized it to have been smallpox. This terrible scourge afflicted the Indians of America all the way to the South Seas as varied European diseases progressively devastated aboriginal "naïve" or unexposed populations. New World human depletion from disease alone is sometimes estimated at up to 95%, the unfortunate Wampanoags suffering 80% or more.

All of the territory of southern Massachusetts from Cape Cod to near Rhode Island was essentially depopulated leaving nearly derelict villages. The Plymouth colonists in 1620 recorded only occasional sightings of individuals or small parties and the remains of numerous abandoned villages. This was because the explorers a short time earlier had been in close enough contact to spread their diseases.

The "Puritan" colonists, indeed, survived for a time on abandoned grain pits of these depopulated villages. Yet in the area surrounding Narragansett Bay, populations seem to have held at least steady. Narragansetts maintained a record of public health comparable with the colonial invaders.

Such were the conditions when Roger Williams arrived in Salem, Massachusetts in 1631. Graduate of Cambridge University, he would have been expected to be part of the ruling clique, but early on he made the blunder of suggesting that perhaps Europeans had moral obligations to pay the natives for seized lands. The general dismay and discomfort of such a suggestion seems to have been immediately felt with the result that he moved to the newer Plymouth Colony, some twenty miles south.

His life seems to be a chronicle of perpetual contentions and he gained enemies wherever he went. They multiplied quickly in Plymouth with the result that he remained there but a short time before moving back to Salem whose governor was John Winthrop (the elder). He dodged charges of heresy and insurrection, to say nothing of his horrendous idea that colonists should actually pay Native Americans for seized lands.

The result was that Roger Williams was sentenced to be transported back to England. Instead, he departed "post haste" to join other exiles in Rhode Island. His relationship with John Winthrop the elder seems mixed. There exist letters between the two suggesting friendship and at least high regard, yet Winthrop must have been a participant in the judgment of transport and threat of death, which that implies.

His seizure had been ordered and if he had been captured by the sheriff sent to apprehend him, that might have been his end. However, he received a warning, apparently from Winthrop himself. He departed rapidly without his family but with a small group of companions.

He apparently reached an area near the tip of Narragansett Bay, decided to settle there, but discovered that he was still within the territory claimed by Massachusetts and subject to seizure yet.

Tensions seemed to be running high when Williams arrived. This was a period of beheadings, hangings, drawings and quarterings, and it was a dangerous time to be independent, outspoken, or heretical in England or English territory, or Spanish, for that matter. Most towns in both America and Europe maintained permanent gibbets and execution grounds. Heads of heretics were impaled on pikes on town walls and pirates were hung in chains near seaports until skeletonized and beyond.

Having escaped with his head from enemies at both Salem and Plymouth, he arrived near the Seekonk River and settled thereafter along another river at a little distance. The headwaters of Narragansett "Bay" (geologically a fjord) divide into a "Y" shape. The eastern arm is Seekonk "River," actually a wide salt estuary with a restricted mouth, while the western arm is so-called Providence River. At that time, the latter widened into a rather large nearly landlocked cove. It was upon the eastern shore of this cove that Roger Williams landed after an initial encounter, and settled where his commemorative museum exists today. The river still flows, but the cove has been filled in.

Williams was 14 weeks in transit of the mere 50-mile distance over what must have been desolate territory of pestilential Wampanoag lands. He attempted to plant a crop before the crossing, which suggests that he was uneasy about moving so far away. He was wavering, wavering about his tentative landing on the opposite shore of the Seekonk. However, his wavering ceased and his destiny for America's future thereafter was established and successful.

Roger Williams is revered in Rhode Island to this day as the founder of the state. Conventional education teaches that he crossed Seekonk River at a site called "Round Rock" or "Slate Rock" into the area of what is now Providence. There he was greeted by several Narragansett Sachems (chiefs) or perhaps guardian warriors with the words "What cheer, Netop." "Netop" was a Narragansett term for "friend." They were so impressed with him that he was able to negotiate the tract of lands of Providence, which are now the Capital city of the State.

This is the dramatic story that is fed to Rhode Island schoolchildren to this day. Not being quite so innocent a schoolboy, I questioned this early on as Seekonk River was well within my wanderings and I knew nothing of any such place where a rock of that description was or could have been. A primary school teacher took up the cudgels to set me straight with the result that she also became mystified. She found that no one she knew could identify the place or the rock. She eventually discovered some reference that the rock was actually some distance south and well out of range for our class to visit in those days before extensive bussing.

I have since been informed, but remain unconvinced, that the alleged first landing place—the rock—in Providence is actually well known. I do not know on what basis my old teacher (bless her) made her remark.

As with so many items of sanitized historical doctrine, the tale makes little sense to me, although reputable historians are reluctant to discount it. For one thing, the rock, if it ever existed, has disappeared, even though just about everything else that Roger Williams touched has become a monument. For another, the tract granted was huge—bigger than the present city and would require a full day's travel just to walk one side of it.

Certainly, some of those Narragansett chieftains dwelt within that large territory. Why would they negotiate their own homes away to a doubtless poverty-stricken exile one jump ahead of the executioner? Moreover, this place was actually near the northern border of Narragansett territory. The tribe's origin, seats of power and population was well south near the sea. The word Narragansett is derived from the name of a revered island (now uncertain which) in one of two estuaries near Point Judith, down on the coast.

It was 50 years before I discovered what might have been the true situation, or at least as plausible a one, which is a bit more prosaic. After much study and observation, my opinion is that Williams saw Indian sentinels on the opposite shore, approached them, and perhaps was greeted there with "What cheer, Netop." He was directed to the second landing out of sight of possible pursuers. The "narrows" of this river are not so wide that a zealous sheriff might not hazard a crossing.

This crossing has always been a factor of interest in Rhode Island, but it was only when I took it under study for this work that I realized just how important it is for all Americans. For it was here at this spot—Roger Williams' Rubicon—that a major upheaval in American political thought became empowered: separation of Church and State.

His life to that time had been as an Englishman attempting to turn the dominion of an established Church toward mercy rather than arbitrary tyranny. Once across the river, he was an American in the New World of the Indian where his success, while not yet assured, was much more probable.

His thinking was doubtless influenced by his belief in the good of the common man, since he came from them. His relationship with the noble and royal class of England apparently while in service to Lord Coke showed him another side of man's nature. According to Mary Lee Settle's biography, he either witnessed in England or became aware of the actual burning at the stake for heresy of a man who had befriended and mentored him as a boy. For someone of even normal sensitivity, this monstrous act, although common in that day, would be life altering and worth risking one's life to correct. It surely must have affected Roger Williams' thinking.

Something seems to have occurred during those 14 weeks of wandering through desolate lands that changed Roger Williams' fortunes dramatically. At that time, there was intimate contact and frequent trade between colonists and

Indians. He was no stranger to either Narragansett or Wampanoag sachems (chiefs), who were themselves regular visitors at Salem and Plymouth.

Roger's humanitarian instincts might have appealed to the natives. His opinion for payment for Indian lands by colonists was doubtless appreciated. Perhaps Mrs. Williams, back in Salem, brought formidable political gifts into the arena. Maybe Roger Williams' intimate connection with a rising faction in England became known, but for whatever reason his arrival at Narragansett Bay was triumphant.

In an amazingly brief time, he had been deeded lands of two huge tracts by two tribes, what are now lands of the City of Providence, the Island "Prudence" in the Bay, and a trading site at Wickford. The Wampanoags ceded an equal area that is now the City of Pawtucket and were influenced to deed the colony of Portsmouth (on the island Rhode Island) to a Mr. William Coddington (first colonial governor) in 1637. Massasoit, the well-known Wampanoag sachem, signed the two Wampanoag deeds.

At this time, European colonists subsisted not too securely only in enclaves directly upon the coastlines. They were enabled to expand westward from Plymouth toward Narragansett Bay because the occupying Wampanoags had been seriously decimated by the 1615 epidemic.

The area of the Narragansett Tribe, on the west side of Narragansett Bay, apparently never extended as far north as Mount Wachusett in Massachusetts about 65 miles away. The narrative of the captive Mrs. Mary Rowlandson (1676) implies strongly that her Narragansett captor (Chief Qanopen/Quinnapin of the Niantic tribe) was familiar with the area but was not within his home territory while at Mount Wachusett. Natick, Nipmuc, and Niantic (still in Rhode Island but to the west) are sub-families of the Narragansetts, speaking the same language and observing the same customs.

The Narragansetts were inveterate travelers and had the ability to travel through and around other tribal territories. Hamonassett, Amagansett, Siasconsett, and Nauset are all Narragansett words. All these places are seaside locales, which in that day were commonly settled only by nautical, warlike cultures like the Narragansetts and Vikings. This is a very different style of living from other Amerinds, and is one of the reasons that I believe it might be a Viking heritage left among these people.

Williams' life, writings, and chronicles have been a subject of intense interest to historians from that day to this. However, no one seems to have addressed the size and placement of the two grants of territory given him by the Narragansetts (the Providence grant) and the Wampanoags (the Pawtucket grant). This transaction gives us some information about the Narragansett philosophy that had unusual differences from other tribes.

The latter was deeded (generally overlooked nowadays) by the same Massasoit whom we celebrate at our Thanksgiving festivities. While closely associated with the Plymouth Colony, in fact Massasoit's home and background was in the Rhode Island area and that grant was of lands north of his own territory.

The Wampanoag grant appears larger than it may have been. Historians say it consisted of the present day City of Pawtucket. Viewing the grants in this manner might give one an impression of considerable political sophistication on the part of the Indians, for that area might have been one of contention and periodic cultural clashes between the two tribes. Giving the lands to the settlers might have seemed a good idea as "buffer" lands between. This might have been the rationale for the facile grant of Aquidneck (Rhode) Island. It contained the proviso that Indians might remain and farm if they agreed to fence their fields. It was necessary to keep out the English sheep and cattle to avoid potential conflicts between Indians and colonists.

The two grants are unique not only by their size but by their inland placement: their heyday was not to come for another hundred years. Newport became the capital of the colony. Williams traveled between Providence, Newport, his granted island Prudence, and his trading post at Cocumcussok (Boy, do I ever like that name, accented on 'Co' and 'cuss').

The Narragansetts are one of the better-known eastern tribes primarily because of a number of cultural advancements and their military strength. Rhode Island's borders are the only state borders established by reality of Indian presence and not by English political dictum. Densely populated, the tribe resided in towns not far apart and in small, crowded dwellings containing as many as 28 persons. Some few larger dwellings contained extended families and large waterside estates. These dwellings were often quite advanced, being cleared of forest in wide tracts by planned fire programs and possessing cultivated, orderly fields. They kept captive wildlife for food and, apparently, sometimes as pets.

Strong cultural markers of the tribe have been observed on Long Island, Martha's Vineyard, and Nantucket. They were also the "bankers" of a wide area over 600 miles, being the manufacturers and regulators of the Native American currency called Wampum or wampum-peague. This was a finely crafted and standardized beadwork of shell fragments from Narragansett Bay used for trading. One early historian says they were one of the two "greatest" tribes on the American Continent, but that was written before the civilizations of Central and South America were discovered.

Part of the payment for Portsmouth included, besides quite modest monetary considerations, the proviso of "four coats." Four coats were delivered but were not appreciated for some reason. They were to be replaced but were unavailable for some time, at which Massasoit conceded the sale without them. Seems like a reasonable enough fellow.

Williams in one place says that he paid "thirty pounds" (sterling) for Providence, while other records say that the grant was from "… love alone of Canonicus (Sachem)."

The presumption that the sachems who deeded the lands to Roger Williams were the same as those who greeted him at Providence may not hold true. Records show that while Williams either landed or closely passed by the landing on the western shore of the Seekonk with some communication, he then sailed or rowed a moderate distance around a southern point ("Fox") and then made his permanent landing at another place. That site, along the shore of a cove in Providence, is now commemorated by a preserved spring and a museum maintained by the United States National Park Service.

However, the deed to the Providence grant shows that it was done "… at Narragansett," some 25 miles nearer the coast near Point Judith, the epicenter of Narragansett Indian Tribal lands. It is more than likely that these negotiations actually occurred at a revered place called Pettaquamscutt Rock, which may be fairly called a "Round Rock." Tradition holds that a popular alternative name for it was "Treaty Rock." Its huge eastern face provides a background for a natural amphitheater ideal for negotiations and ceremonies. It is my opinion that the confusion surrounding the true whereabouts of "Round," or "Slate Rock," supposedly in Providence proper, is from an oversimplification of events by historians.

Williams, with partners named Edward Wilcox and Richard Smith, commenced trading with Narragansetts, and found them by no means so isolated, and more astute than might be expected. While they held their territory secure and were militarily competent, they were friendly and welcoming to the incomers, indeed, showing a considerable political sophistication in where they allowed the new settlements to exist. There is little record that any European was injured or harassed from 1635 when Williams appeared until 1676.

The Narragansetts were adapting well and cooperating with Europeans to the extent that they were entering trades such as locksmiths, farriers, and stonemasons. As well, they were adopting European dress. They owned their own muskets in considerable numbers, which Williams attributed to trade with the

French, some 200 miles distant. One, possibly two, Dutch (from New Amsterdam) trading posts were in existence within their lands.

About the time that Roger Williams arrived, the Narragansetts with their muskets were enlisted as mercenaries against the Pequots of Connecticut, with the result that the Pequot tribe was nearly annihilated. Over the next 50 years, that colony prospered in absence of the decimated Pequots.

Williams appears to have been an accessory in this alliance. He did not think highly of the Pequots. He wrote that at a meeting with some he was barely able to restrain himself and felt he was in the presence of robbers, thieves, and "murtherers." I wonder if the high regard for Williams by the Narragansett sachems may have been appreciation for their success over their hated adversary. The attack upon the Pequots was a joint colonial/Narragansett assault. By 1660, Narragansetts in their turn came under pressure by the Connecticut colonists and were becoming forced to defend themselves against periodic raids into their western preserves.

For many years, Williams maintained his trading post near the tribal center and was on intimate terms with the Narragansetts, even living among and traveling with them. He found them not only astute but also honest and meticulous traders among both themselves and the colonials. Their crafted shell wampum, for a time, had a regulated rate of exchange with English coin. For example, one six-foot long string of shells was worth 10 shillings but a thick belt of shell strings was worth 20 English pounds.

Indian traders used wampum since 200 A.D. The shells with growth rings were mainly found and made along coasts in Connecticut, Massachusetts, New York, and Rhode Island. Mainly two shells were used: one of white (Whelk shells) and one of purple (Quahog shells). The distinctive workmanship and fine drilling made some Wampumpeague more desirable and valuable than others and the Narragansett and Wampanoag were among that category. Many tribes of the Algonquins and Iroquois used wampum among the northeastern United States and up the coast into Canada.

It may have been a number of years before he was well settled in his new grant. "Backwater" Providence developed at a slow pace and did not appear on a navigational chart of 1717. The small settlement of perhaps 20 or 30 domiciles was some 30 miles from the coast. It was some time before the settlement, with the motto "What cheer, Netop," became well enough established that it equaled Newport Colony in influence. Hence, the official name of the area became "State of Rhode Island and Providence Plantations" with the motto "Hope."

Williams seems to have lived for a time in Warwick, not too far from the trading post, but eventually did move to Providence for the latter part of his life. He was instrumental in founding the still extant First Baptist Church in Providence. However, he removed himself from the congregation after a few months. He declared himself a freethinker, or "seeker," despite which he was elected Governor for a time. The impression one might get from readings is that he had a "contentious" personality with frequent spats and conflicts with co-colonists. He even wanted to debate peaceable George Fox of Quaker origin, but Fox beat a strategic retreat from the encounter.

In 1643, Roger Williams traveled to England to promote the cause of establishing the colony as a chartered entity. Williams wrote a dictionary of the Narragansett language on shipboard in transit. It was published in London in 1643 with the title *A Key into the Language of America or An Help to the Language of the Natives in that part of America Called New England*. The cover included these words:

> *Together with brief observations of the customs, manners and worships of the aforesaid natives in peace and war, in life and death on all which are added spiritual observations, general and particular by the author, of chief and special use upon all occasions to all the English inhabiting those parts yet pleasant and profitable to the view of all men.*

In the opening section of this book, he made the following comments about the origin of the Narragansetts from Iceland:

> *Wise and judicious men, with whom I have discoursed, maintain their (Narragansett) origin to be northward from Tartaria: and at my now taking ship, at the Dutch Plantation, it pleased the Dutch Governor, (in some discourse with me about the natives), to draw their line from Iceland, because the name Sackmakan (the name for an Indian Prince about the Dutch) is the name for a Prince in Iceland.*

At this time, England was embroiled in its civil war and Williams seems to have become a close associate with Oliver Cromwell. Indeed, the two are reported to be cousins.

I am uncertain of events while he was there but there were several key battles fought between the parliamentarian "roundheads" and royalists in which it is recorded that New Englanders participated. The battles commenced with Grantham, May 13, 1643, Gainsborough, July 28, 1643, Winceby, October 11,

1643, and the Battle of Marsten Moor, July 2, 1644. In my opinion, one of these New Englanders might well have been Roger Williams. Not only that, I also strongly suspect that perhaps several of his Narragansett friends were along with him.

For one thing, his dictionary bears all the hallmarks of much closer and immediate collaboration than might have been done on shipboard by memory. It is true that it could have been written from memory, but 1643 was only eight years after his residence among them and the dictionary seems much deeper and more extensive than this experience might explain.

Secondly, Cromwell won the battle at Marsten Moor thanks to an unorthodox and dashing cavalry "end run" around a flank into the rear of the enemy, conventionally arrayed in the set-piece manner of the day. This type of maneuver was more typical (except for the horses) of Indian combat than European. It was also the type of thing that, if successful, might merit recognition or a grand reward. American Indians were beginning to appear in European streets and courts. There is good reason to think, though apparently unrecorded, that one or two perhaps were there with Williams, their European benefactor.

The result of these battles was the loss of the royalist cause and the end of King Charles I to the popular commoner, Oliver Cromwell. I do not know how long Williams was in England on this visit. Later he returned to England where King Charles I's son (Charles II) had been re-established on the throne but not until 1660. Williams is reported to have traversed the Atlantic Ocean five times. The result of these adventures was that he or an associate named Clarke, or the two together, returned to America in 1663 with a Royal Charter for what is now the State of Rhode Island.

That charter specifies that citizens of the colony enjoy religious freedom, which the polytheist and independent-minded settlers had already established there. Furthermore, that freedom was extended to the Narragansett Tribe—staunchly pagan, and neither Biblical nor Christian!

This is how the separation of church and state originated in America. It appears to have been the result of initial Indian tolerance of invaders, followed by a reciprocal respect, tinged perhaps by fear of the dominant Aborigines. It is due, also, to Roger Williams' humanitarian instincts, to a new and insecure King of England whose headless father's body was only recently buried; and a considerable upheaval and religious rebellion throughout the realm.

Near the 1660's, there seems to have been a notable movement of Narragansetts selling large tracts of their homelands to Rhode Island and other colonists. Roger Williams, in his capacity as elder statesman and Governor, together with the Rhode Island legislature at Newport, tried to prevent this. This seems a strange reversal for Williams whose prosperity contributed to the first native grant.

My own unsubstantiated idea is this. Just before the Royal Charter was delivered (about 1663), both Massachusetts and Connecticut colonies had aggressively attempted to seize Wampanoag lands westward to Narragansett Bay and Narragansett lands eastward to the same waterway. Connecticut (actually Massachusetts) Colony policy seems to have pressured the Narragansetts to sustain periodic military attacks upon native towns and persons at the western side of Narragansett territory. It seems plausible that the sale of lands within would make for good alliances against these outside pressures.

The Narragansett victory over the Pequots thus became a Pyrrhic one, although this was not yet recognized.

PART V
▼

NARRAGANSETT DIFFERENCES

We detail previously unexamined recordings of these people and find them unique among all other American Indian tribes. We introduce recent archaeological evidence that this uniqueness has its roots in an archaic incursion with dating coincidental with that of the Vinland voyages.

Physical Differences

The Amerind tribe of Narragansetts was found to be different from other tribes in a number of unique traits. Most Amerinds are descendents of those who traversed the narrow Bering straits eons ago. However, genetic studies that will be described herein show that the Narragansetts must have been a consolidation of two hereditary lines of descent. Many have noted that there were and are different physical characteristics from other Amerinds in the area. Even more importantly, these Narragansett Amerinds had developed immunities to European diseases which were also noted by many, have now been proven, and will be developed later.

When Giovanni da Verrazzano arrived in 1524, he noted that there were fires burning all along the shore of Block Island (his Luiza), likely a signal for trade, but he could not land. At Narragansett Bay, the Indians off Point Judith approached in canoes for trade.

He encountered Narragansett spokesmen and recorded that they were taller, that some were white to tawny and described their facial features as "fine." A rare Native American can be so described. Most male Amerinds age by developing a sort of weathered, craggy appearance. Northern Europeans and a high proportion of southern are typically fine-featured with the trait being more common toward the north.

A strong inference in the early records of the Plymouth Colony indicates that some of the Indians seen were bearded. By the middle of the nineteenth century, beards were thought more typical of Narragansetts. References to the Wampanoags, we must recall, were of the western groupings of the Wampanoag tribe, adjacent to Narragansett Bay, as the eastern sectors were all decimated and mainly occupied by the incoming settlers.

Indian hair is universally coarse and straight. It is not always black. In fact, on the Southwestern (Arizona) reservations, the more typical color is a very dark brunette, often with lighter brown highlights, which makes the hair appear an auburn color. That was the way Gudrid Thorbjornsdottir described a Skraeling in the sagas.

Facial hair on Amerinds is nearly non-existent, being very "wispy" if it occurs at all. Many simply pluck out the few strands. European hair is commonly of much finer texture than Native American and is, of course, of many shades and variations. Lighter color seems the initial casualty of such intermarriage. A few generations and it is gone, jet black predominating. Nevertheless, many of the Native Americans in the east today are quite light in hair color.

Regarding physical characteristics, blue eyes and blond hair are recessive genes and would not be expected to be sustained if Native Americans interbreed with Europeans. However, we do have a hint about the texture of the hair of Narragansetts when seen in 1524 and the 1600s, if not the color, by both eyewitness descriptions and an oil painting.

A painting of Ninigret hangs in the Museum of Art in Rhode Island School of Design in Providence. I should mention that the Museum says it is Ninigret II, the son of Ninigret, but the family of Winthrop who personally knew Ninigret donated it. Most scholars believe it is Ninigret, sachem of the Niantics, whom John Winthrop, Jr., knew and admired.

An artist painted it around 1638 in honor of the Narragansett/Pequot War. Winthrop was one of those who encouraged and actually allied colonial troops on the Pequots. He was personally known to Ninigret, a "next door neighbor," and had both the means and the motive to commission this work as a record of the victory. (If it were from a later period, Indians dressed in English, French or Dutch clothing.)

It clearly shows a "loft" of finer hair than is encountered anywhere else among Indians, rather than coarse and straight.

Ninigret was a famous Narragansett (Niantic) Chieftain, whose ancient home site is still in existence in Charlestown, Rhode Island. He died or was killed in 1676. His skeleton was exhumed in 1853 and he was found to have been six feet two inches tall when average Englishmen were but 5'3". In this painting of his son (the next best thing to a photograph) are several anomalies described above. It can also be compared with descriptions of Hop's "small" Skraelings from the Vinland sagas.

Additionally, the painting of the Chieftain Ninigret shows at least four genetic anomalies coincidental with an interchange of genes: "loft" of fine hair, stature, very light skin, and the size of his eyes.

Ninigret's hair over his right shoulder demonstrates what Verrazzano meant in 1524 by "flowing." It is atypical of ordinary Amerinds, which almost invariably are depicted with hair that hangs straight without such a "loft" as we see here. The artist appears to have treated the customarily worn cape with some uncertainty.

Interestingly, Narragansetts also wore a non-utilitarian and stylized cape, a habit not known elsewhere among Amerinds. Williams said they seldom used it, not even as a blanket, but women used it for modesty, maybe when seated or squatting. To me this is highly significant and a natural holdover from ancient customs.

In the painting, the cape seems to have been deliberately brought part way forward to show it. It is not unlike Viking capes. Viking capes were a standard item of dress, not only present but with a stylized manner of wear—men wore a short cape clasped at the right shoulder and women wore a longer, ankle-length cape clasped with a characteristic double brooch at the neck. These were a uniform dress for Vikings, as was the simpler one for the Narragansetts.

Roger Williams said that the Narragansett cape was worn directly down the back and here it seems to have been brought around for better display. The material seems uncertain and may even have been some sort of lace. His headband and necklace are of wampum (peague). The staff he holds may be a symbol of his position. There is something in his waistband, which could be a knife. His footwear appears crude and may be a form of legging used for digging clams—which is best done on the knees.

He stands upon what appears to be a shingle, which drops off to some waterside beach. It resembles the terrain on the north bank of Narrow River, which demonstrates the unique proximity of forest to waterside, which is stipulated in the sagas.

Recent reproductions of the same painting in two other modern historical works both show the skin color as much darker than the original painting, apparently to match imagined comparisons with neighboring tribes. Mohawks (100 miles west) were sometimes as dark as Negroes were. Roger Williams observed that Narragansett skin color was white at birth and only darkened with sun exposure and "annoyntings" (tattoos). Also in 1524, Verrazzano described the skin color as "light to olive."

That merger of races (Vikings and Amerinds) I hypothesize to have occurred a thousand years ago in the Pettaquamscutt River basin. The Narragansetts' coloration, hair texture, stature (6'2"), and possibly Chief Ninigret's lips suggest Caucasian descent. His enormous eyes recall the strong descriptions the Norse sagas attributed to those natives they knew in Vinland.

Early descriptions of Narragansett white skin, atypical muscular and skeletal structure, aberrant hair distinctions, and unique social mores can only support the deduction that the two races mixed further.

Roger Williams and associates in the 1636 colony of Rhode Island guessed that Narragansett origins had been in Iceland. As far as I have been able to determine, neither Williams nor those associates connected this anomaly with the Vinland sagas, their thought being simply from observation of the genetic and anthropological peculiarities of the group and awareness of geography. This remained a lively issue until the middle of the nineteenth century but is largely forgotten today.

The present day Township of Narragansett is a narrow band along the West Coast of lower Narragansett Bay. The borderline between the townships of Narragansett and adjoining South Kingstown is directly up the channel of the Pettaquamscutt River. The Narragansett township entity has had the name since colonial times, thus indicating Pettaquamscutt as at, or near, the main locus of the Narragansett Tribe.

The finding of a Narragansett cemetery (site RI 1000) with Ninigret and his family and other Amerinds has generated much interest because of cultural indications and evidence of resistance to diseases found on skeletons.

Rhode Island state archaeologist and preservation specialist, Paul A. Robinson, and scientists wrote about these Amerinds in the article "Preliminary Biocultural Interpretations from a Seventeenth-Century Narragansett Indian Cemetery in Rhode Island," in *Cultures in Contact: The European Impact on Native Culture Institutions in Eastern North America, A.D. 1000–1800* in 1985. Robinson has also written another helpful article: "A Narragansett History from 1000 B.P. to the Present," in *Enduring Traditions: The Native Peoples of New England*, edited by Laurie Weinstein *et al.* in 1994.

More detail on the RI 1000 site can be found in William Turnbaugh's *The Material Culture of RI 1000, a Mid-17th Century Burial Site in North Kingstown*, published by the University of Rhode Island in 1984.

It is not enough to describe simply physical characteristics that suggest interbreeding with other races. Vilijhalmur Stefansson knew that because he endured much criticism when he told reporters about having seen "blond Eskimos" upon returning from life with Inuits in 1912. He soon learned that appearance alone was insufficient to conclude racial integration.

While living among Canadian Eskimos in the Coronation Gulf area, Stefansson had written in his diary:

> *There are three men here whose beards are almost the color of mine, and who look like typical Scandinavians ... The faces and proportions of the body remind of 'stocky,' sunburned, but naturally fair Scandinavians.*

Newspapers called these people "Blond Eskimos." Stefansson's observations suggested that Norsemen had interbred with Eskimos. His assertions generated much criticism. He responded to that criticism with an article entitled "The Blond Eskimos" in the December issue of the 1912 *National Geographic Magazine*, excerpted here:

> *In 1000 A.D., Erick's son, Leif, discovered the mainland of North America; and there followed during the next 300 years many voyages to Baffin Island, Labrador, Newfoundland, and the other parts of North America, perhaps as far south as where New York now stands ...*
>
> *The "Blond" Eskimos, among whom I lived a year, are interesting for many reasons, only one of which is the light complexions of some of them ...*

He wrote the fascinating 1913 book *My Life with the Eskimo*. He had to defend himself against others who, in some cases, searched for and could not find blond Eskimos. Nevertheless, that fueled a search to discover how the Vikings might have gotten through the Northwest Passage from Greenland to Victoria, an intriguing quest in its own right.

Returning to the Narragansetts, over many years in pursuit of this whole subject, I gradually realized that there were many differences concerning their social structure and even their language. The earliest colonial settlers in Rhode Island kept notes of genetic, anthropological, social, and linguistic anomalies unique and at variance to even immediate neighbors. So striking were the observations that some English and Dutch speculated that Narragansett forebears had originated in Iceland, according to Roger Williams.

LANGUAGE DIFFERENCES

All languages are made up from a combination of several other languages. For example, English is made up of Old Norse, German, Latin, and Norman French according to *The Story of English* by Robert McCrum, *et al.* Currently, there are some 6,000 to 6,500 languages in the world and they are dying off at the rate of one every two weeks. There were once around 1,000 Native American languages, but they have dwindled to some 143 languages, according to linguist Johanna Nichols.

Some specialists in language aver that it is as powerful an indicator of man's migrations as any other science. Famous linguist Merritt Ruhlen has detected anomalies in the languages of Amerinds indicating at least some transmission of European language into American native language. See his 1996 book, *The Origin of Human Languages*.

How did this infusion of language occur? It seems near certain that this particular intermarriage took place from invaders siring progeny with females taken, purchased, or seduced from aborigine groups. In the original group, a class of infants were probably raised in households speaking their mother's tongue, being strongly influenced by their father's social and structural mores, and perhaps picking up a word or two where their mother's tongue would not suit.

We are indebted, again, to Roger Williams who not only spoke the Narragansett language but also actually wrote a dictionary of it supplemented by the work of Reider Thorbjorn Sherwin who wrote an analysis of Algonquin languages. *The Viking and the Red Man: Old Norse Origin of the Algonquin Language* was published from 1940 to 1956. He found some 15,000 words in the Algonquin language that he believed had Old Norse roots.

Sherwin claimed that the general frequency of words in common throughout all Algonquian enabled him to detect not only several incursions of Norsemen at varied latitudes, but also he added that he could actually date them. He remarked that among that sequence, Mic-Mac of Maine/Nova Scotia appeared the most recent and Narragansett the oldest of incursions because of the expected systematic corruption and loss of words over time.

Roger Williams' original thought was that Narragansett language did not seem to him to have many, if any, words from Europe in it. The Dutch Governor of New Amsterdam who identified the word "Sackmackan," which meant "prince" in Narragansett and Old Norse, enlightened him. It probably evolved from that to "Sagamore" (chief or king) or "Sachem," (chief). The Indian was nothing if not a free spirit in regards to pronunciation. That a Dutch Governor (Peter Stuyvesant or Peter Minuit) had enough awareness of the incidence of "prince" in both languages that an intimacy of both peoples is implied, but Williams spoke Dutch to the skill level that he was, for a time, a tutor of the language to Milton.

Netherlanders maintained the trading post at "Fort Ninigret," so they were intimate with the language, Ninigret the chief, and the Narragansett people. In fact, the Germanic languages include German, Dutch, Old Norse, English, Icelandic, and Swedish.

From Williams' point of view, we might wonder why he should think there should be any words in common in the first place. But we already know from his other writings that almost from his first visit to Providence in 1635 he had become a firm believer in the idea that Narragansett origins had been in Europe, so it would seem the mere suggestion by the New Amsterdam official would suffice to point him toward Iceland as the nation of choice.

Let us examine the language as Sherwin and Williams have done. An insightful item of the language is that the letter "J" was virtually absent. Well, by yumpin' yiminy, if my Dad spoke that way, I think I might pick it up myself! This particular sound is noted as a major impediment to Scandinavian immigrants as they move into English-speaking lands.

A good local example of this peculiarity is in the modern vernacular "Yankee," which designates specifically a New Englander as opposed to any American. "Yankee" is derived from the New Amsterdam Dutch who referred to New Englanders with the pejorative "Jon Kees" ("Jon Cheese," an ignoramus or a rustic and, of course, pronounced "Yon Kees").

Need I remind the reader that the Dutch are every bit the descendants of Vikings just as Scandinavians are? The formerly impalpable Rhine delta "Netherlands" are, in fact, the nether lands and sometime provinces of Viking Denmark. Indeed, much of their language is derived from the same Old Norse of the Vikings and modern Icelanders, and they are every bit the followers of seagoing ways. They are the experts at utilizing the seaside environment, having manufactured their own country, and are practiced in ship salvage operations. I often wonder at the early founding and development of New Amsterdam, which is located at the mouth of a great waterway matching trade opportunities of their own homeland.

Another is the frequency of the letter "N" in major Narragansett identifiers including its own name. Vikings seem to have been partial toward that sound/letter as it occurred in their own favored "Norse," "Norsk," "Nordic," "Norman," eventually "Norway." As identifiers here, we can find the words like "Natick," "Niantic," and "Nipmuc" as sub-tribes. These might reasonably be followed by "Nope" (Martha's Vineyard) and "Nomans" Land Island.

By the way, let me explain about the name of that island. It is widely believed that "Nomans" is, indeed, a corruption of "Norman's" and the capital "M" being a later addition from a belief that it was meant to mean "Noman's Land;" but the spelling has always left no space between "No" and "man" which leaves some confusion. Modern nautical charts capitalize it "NOMANS LAND" while the "Coasting Pilot" of Captain Cyprian Southack of 1717 has it as "Nomans land Island," *(sic.)* That it might be Norse in origin is not a far-fetched idea, whether the custom originated a thousand years ago or 400 years ago. It may have been so-named because of a runestone still present along its southern shore, which will be described elsewhere in these pages as well as the Internet.

My belief is that Martha's Vineyard as Straumney must certainly have been as important a Norse site as Pettaquamscutt. Nomans Land Island is certainly a most ideal and typical Viking site.

A name at the "elbow" of Cape Cod is "Nauset." which is a Narragansett place name, quite possibly meaning "Nau" equaling "North," and is phonetically quite similar to "Nor." The Elizabeth Island chain just west of Martha's Vineyard was called by the natives "Nashanow" or "Naushanow" with the largest being "Naushon," with others being named Na(u)shawena, Nonamessett, and the adjacent mainland peninsula, Nobska. Other islands, or places upon them, were named Saconesett, Nanomessett, and Peshchamesett, all Narragansett names, the latter with added significance, which will be addressed below.

The entire North Atlantic was the domain of Vikings for over 500 years. They owned it completely. Any ship sighted offshore from about 800 A.D. to 1300 A.D. was certain to be Norse. Indeed, the entire world's oceans were theirs if they had the mind and time to go where the winds might carry them. Recorded voyages from Iceland and Greenland span along the American coastline down to Brazil equivalent to the distance Vikings traversed to Eastern Europe and Asia. The nautical outlook of Narragansetts was a close comparison to Viking seamen's and this is a trait we might expect to survive and perpetuate in such a mixed society.

Amerind words are composed by chaining together descriptive syllables to make up a word to describe a thing or place. Old Norse syllables were significant and the Narragansett language did the same. A rather comic use by Narragansetts was to name newly introduced animals by their sound such as; "honk-honk-suck" for domesticated geese, "oink-oink-suck" for pig, and "Neigh-neigh-suck" for horses. Thus, the syllable "suck" must mean a living being, an animal or bird or perhaps simply "food."

Place names were treated the same and a plethora of those names are still very much in existence all through New England, while the Indian meaning of the chain is lost. They might refer to a place as, for example, "place where water flows over the rock." If we can examine enough names, we might be able to decode some of the language. Chaining syllables is a trait of most Teutonic peoples such as German, Old Norse, and Icelandic.

Early on in my correspondence, an elderly resident of Reykjavik, Iceland, named Skulli Olafsson, contacted me. He had been a Vinland scholar for most of his many years, knew myriad details of the lost land, and was eager to share his wealth of lore as he neared the end of his days. He informed me in his small and crabbed hand (appearing to have suffered great pain in the process) that "many Icelandic scholars believed that Vinland had been in Rhode Island," because "Narragansett" could be parsed in Icelandic so: NARR (meaning "fool" or "deceiver"), GAN ("going" meaning "trip" or "voyage"), and SETT (in general simply meaning "place"). This suggested the phrase "a fool's errand." "Sett" can also mean a settlement or an extended farmstead. Thus, in Icelandic or at least Old Norse, Narragansett is a recognizable word meaning "fool's trip settlement" or a place/settlement to which fools or deceivers traveled or resided.

Does this fit what we know of Vinland? Yes, it does, because those who stayed in Greenland probably thought the Viking adventures to distant lands such as Vinland to be foolish. On the other hand, perhaps they thought it foolish to expend excessive labor to round Cape Cod when the same commodities could be found without the effort. Recall, also, the final episode of the Vinland sagas where Freydis was condemned as much for deceit in her attempt to conceal murder as for the murders themselves. Therefore, Vinland voyagers might have been considered fools, liars, and even braggarts about grapes, wheat, and other supposed treasures.

As it happens, the suffix "sett" has the identical meaning in Narragansett as it did in Old Norse. In modern Icelandic, it apparently means "place", "homestead," or "settlement." Curiously, I was also informed by the late learned Viking scholar, Dr. Albert G. Hahn, that it also seems to have been the origin of the suffix "ster" in English place names from those days of old when Vikings dominated Ireland and the midlands of England, as in Worcestershire or Lancaster, etc.

Let us compare what archaic Narragansetts themselves had to say of their own name. In general, Amerind tribes referred to themselves egocentrically, with their identification being translated into "We," "Us," "The Human Beings," and "The People." This almost universal attitude of superiority to their neighbors helps explain their reluctance to form alliances and their sadistic behavior toward prisoners of war so common among Amerinds. Williams wrote this about names:

> *I cannot observe that they ever had (before the coming of the English, French, or Dutch amongst them) any names to difference themselves from strangers, for they knew none; but two sorts of names they had, and have amongst themselves.*
>
> *First, generall, belonging to all natives, as Ninnuock, Ninnimissinnuwock, Eniskeetompauwog, which signifies Men, Folke, or People.*
>
> *Secondly, particular names, peculiar to several nations, of them amongst themselves, as, Ninnegganeuck, Masschuseuck, Cawasumseuck, Cowweseuck, Quinticoock, Qunnipieuck, Pequtoog, etc.*

My guess is that translated into modern place names, since these are tribes or nations, these would be Ninnegansett, Massachusetts, Cawasumsett, and Cowesett (I'm sure of that one), Quinticook, Quinipiac, and Pettacook.

"Narragansett" is usually translated to mean "People of the Small Point" and this seems to be the belief of modern descendents. However, their language declined as the tribe dissolved after 1676 and became lost to themselves. It was preserved mainly in Roger Williams' 1643 book. According to Williams, the name had no meaning to them. Think of it! Such a complex word appears to have been lost to their oral memory by 1635. Yet it appears in the language of Icelanders whose memory was written and therefore much longer!

Now, well-educated Williams spoke it day to day while he lived among and traded with the Narragansetts for over thirty years. He found it to be the name of a small island, which they revered, but to which the linguistic meaning had been lost to time. Modern historians cannot ascertain just which island this may have been from the indeterminate descriptions left by Williams.

Most have looked into adjacent Point Judith Pond, which has six small islands. Actually, that is a salt-water estuary, as is Narrow River, the very short entry to Pettaquamscutt River. However, the descriptions are even more potent for "Gooseberry Island" within Pettaquamscutt Cove. "Cove" is another misnomer for this lake, freshwater then, and brackish salt now. Williams said he had approached it to within a "pole" (a "rod" or about 16 ½ feet).

Interestingly, Gooseberry Island in Pettaquamscutt "Cove" has the narrowest strait of all the possible islands we can consider. I would judge it at about 25 feet from the nearest land. A running dive would bring an adventurer to the island itself, but apparently, Williams was not in that particular mood on that day, or perhaps he was too cold. Enthusiastic campers still popularly access it. The first time I landed, there was a small campfire still smoldering among the large rocks—bedrock, which is why it is so durable.

Therefore, it seems plausible that Gooseberry Island was a significant place in Narragansett culture and possibly the origin of their name. An aerial view of it reminds one of an arrowhead—even complete with a notch at the base. It is so small that even those who have not visualized the bird's eye view can realize this viewpoint.

It must be considered that the Vinland sagas also define a "point" at the settlement of Hop. A "point" in modern nautical parlance is the same as "headland," but usually smaller. Headlands usually separate inlets of the seas called "bays" while a "point" might not necessarily do that—simply extending out into the water. This particular point and its distinctive overlook so well described in the sagas also still exists a hundred yards or so north of the Gooseberry Island.

The word "Hop" in Old Norse had a very precise definition: "a fresh water lake into which salt water flowed at high tide." Only a small percentage of estuaries fit this description. It implies certain geological conditions of a river mouth, which restrain tidal flow until near high tide. Often the sudden appearance of salt water is noted as a "tidal bore"—a more or less distinctive incoming wave. Narrow River entry does so fit and quite precisely.

The sagas' description of Hop said that the river flowed from north to south into a lake, and then into the sea. This also is an unusual topographical description. However, Pettaquamscutt River flowing into what is called a "cove" (misnomer), which is itself exited by an entirely different river, fits this description perfectly. The second river is a mere ¾-mile long (a ten-minute row or walk) and is called "Narrow River." It is so distinct that no one ever tried to connect it to Pettaquamscutt River. The unusual configuration of the waterways is exactly as the sagas describe Hop. A lake as a fresh water feature is rarely located so near the ocean. More typically, "lakes" are situated nearer headwaters and upper reaches of a river.

While most names with the suffix "sett" are in or near Narragansett Bay, there are clusters of them near Plymouth and Barnstable Harbors, some five or six each and some are duplicated several times. A river near Plymouth was named "Pawtuxet" and another river of the same name enters Narragansett Bay near Providence. The phonetically identical suffix appears as "set," "sett," "sit," and "xet." Usually the names are accented on the syllable prior to "sett." Thus, we find "Paw-tux'-et," "Quon'-set," "Ponn-a-gann'-sett" and a variant "Som'-er-set."

Here is a further sampling of names: Quidnesset, Aquidnessitt, Chipawanokset, Tippecansett, Mattapoisett, Montponsett, Asawompset, Touisett, Towesett (pronunciation similar but separate places), Pocassett, Saconset, Chacapacasett, Coonampset, Cowesett, Ponnagansett, Aponagansett, Pasquisett, Cohaset, Onset, Cochesett, Siasconsett, Segregansett, Acoaxet, Neponset, Hookset, Wapanayset, Copasnetuxet, Romegansett, Wapanayset, and Wannamoiset. Not all of these are within Narragansett Bay, but they center inside Narragansett Bay or along areas of Narragansett visitation, if not dominance.

Two streets, which conjoin at what had been water's edge at Providence River in the City of Providence, are Westminster and Weybosset, the former derived from early English colonists' nostalgia and the latter from the Narragansett settlement that once existed there. Indeed, we can envision a period of idyllic co-existence of two cultures at this waterway (Indian "Moshassuck" River/Cove). Providence was on the east side of the river and cove while the Indian "Weybosset" village was a mere few hundred feet opposite on the west. In going about their respective businesses, the populations were not only within sight, they were often within reasonable conversational distance.

As "civilization" moved west, the main division between the two cultures was that the invaders came and settled invariably along rivers, while Amerind cultures approached rivers for water but seldom resided there.

Around Narragansett Bay, the residual "sett" sites seem to vary in size from 20 to 80 acres, as Verrazzano described, "… large enough for an army to maneuver." Only a few are located in inland areas such as "Neponsett" and "Wachusett" but well within Narragansett dominance. Wachusett may be derived from a Martha's Vineyard Island hill called "Wachusade," or perhaps from "Wautchaug" (Watch Hill) in western Rhode Island.

Siasconsett is on Nantucket Island and shares extreme easterly placement with Nauset—both exposed sea vista locations atypical of Native Americans dwelling styles. Siasconsett is a Narragansett word translated to "place of many bones" such as whalebones from strandings. (Remember the stranded whale in the sagas.)

At least one "sett," Mettatuxet, is located along Pettaquamscutt River itself, and several are duplicated. "Ponneganset" and "Apponogansett" both seem to occur in two separate places and perhaps three. Note the "gan" syllable, which is quite common. Narragansett Bay is large enough that travel by boat is best described as a trip, taking a day or more in transit. Pawtuxet is a good one-day boat trip north of Pettaquamscutt. Causumpset to Rumstick Point is another similar distance. In fact, while the New England estuary is slightly smaller, it is remarkably similar to Oslofjord in Norway and no one ever thought to identify *that* as a "bay."

The "sett" suffix is therefore transposed in the two languages with identical meanings of "place," "settlement," or "homestead."

There seem to have been some 50 or 60 of these "sett" suffix place names. Some have been lost to time and exist only on archaic maps and in colonial references, but quite a number still exist. Most are located at waterside locales, less often along interior rivers off the Bay. The most distant seems to have been Hammonasett, on Long Island Sound near the Connecticut River, some 60 miles west, and the furthest away on the opposite course in New Hampshire. The furthest inland is a small, sugarloaf-shaped Mount Wachusett, some 50 miles northwest in Massachusetts. Moreover, as it happens, there is much evidence that it might also have been a revered Narragansett site.

One of the past Indian settlements, on the east side of the Bay and therefore Wampanoag, was Causumset. It is now known as "Rumstick Point" near present Warren, Rhode Island. As Causumset, it is nearly forgotten and little thought has been given to how it got the unusual old name of Rumstick Point. Many suppose that it has something to do with the manufacture and/or handling of a liquor of the same name. Rum, a derivative of sugar cane, was widely imported from the Caribbean in colonial days and was used as a food preservative as well as an intoxicant. Being a distillate from huge quantities of sugar cane, it was always distilled with fuel obtained from the spent stalks, near where cane grew in Cuba and the West Indies. However, that does not seem to be the origin of the name.

It just so happens that there is a somewhat comparable word to "rumstick" in Old Norse—"Romstovkl" or something like that. It means "room—stick (or stave)." The spaces between the ribs of Norse ships were also called "rooms" and each one had oar ports on either side of the ship. Therefore, if it did derive from Old Norse, it might mean simply "oar," although there are varied other meanings of "roomstick."

Therefore, we find these few traces in language to suggest connections with Old Norse. It would be convenient if we could find that the language of the Narragansetts was identical or nearly so with Old Norse, but the realities of life and the vicissitudes of time do not permit it. As any linguist will tell you, languages change very quickly until they become written languages, and then their change still proceeds but at a somewhat slower pace. Nevertheless, there is something more than nothing, and what we do have is exceedingly supportive and insightful. The opinions of two scholars (Sherwin and Williams) and their contacts with the tribe cannot be ignored.

In combination with the powerful anthropological and social indicators, it does contribute to the argument that the Narragansetts may have had archaic contact with Old Norse speaking peoples. The fact that Roger Williams accepted the word "Sackmackan" to clench his opinion of European origins to Iceland is enormously significant.

The value of the "sett" suffix lies not so much as a proposal of a simple linguistic theory, but in its context of variance in life styles. These sites are unusual in Amerind culture but so typical of Norsemen; seemingly more advanced than the equally nautical Northwest Pacific Indians.

Since I had been raised in the area and knew some of these place names, when I received the letter from Mr. Olafsson, I was immediately struck by the presence of other sites with the suffix "sett" and a few including the middle syllable "gan." Some learned correspondents informed me of place names in the British Isles where it had been transposed to "ster" but actually stemmed from the same Old Norse root suffix "sett," or "staedr."

One scholar in England retorted that "ster" was, in reality, from Old English identifying a monastery. I could have debated that with the viewpoint that Old Norse is older than Old English and that the suffix even as a monastery might still be a valid holdover from the more ancient source.

The existence of the word "prince" itself is an anomaly for Amerind social structure. Almost never did Native Americans consider this type of hierarchy—it seems unique to Narragansetts. Indeed the difference in social attitudes toward leadership cannot be more evident than by comparison in the two groups at identical latitudes on opposite sides of the continent. Among Northwest Pacific Coast Amerinds, there is an impoverishment of the chieftains compared to the dominance (such as the rights of execution) of the Narragansett Sackmackans.

Another interesting language anomaly appears in a game. Narragansetts and Wampanoags enjoyed what appears to have been an immensely popular gambling game they called "Hub." William Woods, an English poet and pamphleteer who settled in Massachusetts, described it in his 1634 book *New England's Prospect*.

Play was with five disks, white on one side and black on the other, which were placed in a shallow tray of some 6-8 inches across, itself set in a shallow depression in the earth. A player agitated the tray in such a way as to make the pieces jump while at the same time fanning his free hand above it in an attempt to influence the fall of the disks. While trying to influence the fall of the disks, the player would shout "hub hub hub" which means, "come come come." The resulting fall by proportion of the colors was an object for the game of chance. Precisely what the result was supposed to be, Mr. Woods did not say but I would presume that the object was to get all one color or the other of all the pieces and perhaps one color would be superior to the other.

This was played with many spectators all gambling and when play was in progress the entire party would shout "hub-hub-hub" so loudly that Woods said you could hear it at a quarter mile distance. Oxford dictionary derives "hubbub" from Scots or Gaelic "hoib" which, however, was only a seldom used cry and not necessarily repeated. It seems far more likely that the English "hubbub" is derived from New England Narragansetts/Wampanoags. This idea is supported by Oxford's dating of "about 1645" as well as a usage example of near the same date by one Captain Hawkins who said his crew used it as a battle cry in attacking Polynesians, adding also that this was "just like the Indians."

Certainly, the linguistic approach might seem feeble, but other supporting evidence extends the argument that Narragansett Amerinds truly are descendants of the Vinland voyagers.

For further details on language differences, see some of the multi-volume work of Reider T. Sherwin's, *The Viking and the Red Man: The Old Norse Origin of the Algonquin Language*.

For a discussion of how the Vikings changed Old English, see John H. MacWhorter's *The Power of Babel: A Natural History of Language*, published in New York by W. H. Freeman in 2001. He describes how the Vikings, who spoke Old Norse, tackled Old English as adults when they came to the British Isles to trade, raid, and settle. Adults can never learn a language as perfectly as children can. They did not know whether a table was male, female, or neuter. The stripped-down rendition of Old English by Vikings streamlined the language leaving out gender endings. English assigns no gender to "table" or any other word except relating to people or animals.

The Vikings also dispensed with reflexives. Therefore, in English, we do not say "I sit myself down" or "I hurry myself"—we say "I sit down" or "I hurry." MacWhorter also described the hordes of new words which entered Old English from Old Norse beginning about the late 8th century. As he says, "language changes leave footprints" and the Vikings have forever changed our English language.

NUMBER DIFFERENCES

Narragansetts were the primary manufacturers of Wampum (correctly "Wampumpeague") which was a respected medium of exchange for over 600 miles, far and away distant from their limited 20 by 30 mile square homeland. This medium was made from shells extracted from Narragansett Bay. It was so refined as to have had for a time rates of exchange with English money. It had standardized widths and perhaps lengths as well, but in barter it could be, and was, cut to varied lengths. Like their boats, it was made with "great artifice" and consisted of black and white shell fragments.

Division of labor to this degree was unusual in Amerind society elsewhere. Most Indian cultures aimed rather at self-reliance by the individual. Apaches, for example, were known to have made all their own moccasins (that also is a Narragansett word, by the way). In later history and elsewhere wampum was made from beads of many colors, also quite decorous, but the beads were mainly of glass and from European sources.

As a toolmaker, I am intrigued by the technical challenge of drilling numerous small holes in the shell material. As an amateur seaman, I am equally intrigued by the challenge of knotting the strands in such a way as to permit cutting the whole without disintegrating the parts. The black shells were, I think, from a very common snail resident in huge numbers on the black mud of Narragansett Bay. Most historians claim the black was from a part of the hard shell clam called, locally, "quahogs." As a resident of the area, I have never seen a quahog that had other than a very thin and purple section of the interior of the shell—not black at all or thick enough for the purpose. I believe that the wampum shell was from another species of shell sea life.

Wampum had many uses and one of them comes from a story by a preacher's wife, Mary Rowlandson. A captive of King Philip's War in 1676, Mrs. Rowlandson was held and ransomed for 20 pounds Sterling. During her 11 weeks of captivity, she narrated an incident that happened when Chief Quanopen/Quinnapin came to visit along with his wife. It seems to have been an amiable enough occasion and the two conversed at length over several days. However, one evening, for some reason, he became annoyed with his spouse and after some words, chased her out of the wigwam and around the structure. Mrs. Rowlandson remarked that during this interchange, his wampum, often worn around the neck, was "… jangling around his ankles," which gives us some insight as to what wampum was like, how it was worn, his status, which must have been noble. She also described this apparently termagant wife, saying, with some sense of cattiness, that she spent hours preparing herself for the day, expending as much or more effort than the greatest of English Dames.

Permit me to add just a word about King Philip's War, which will be discussed later. Massasoit, Wampanoag of Narragansett Bay, had two sons who were called by the English "Alexander" (after the great one) and "King Philip" (after the Macedonian father of Alexander.) I think his Indian name was Metacomet. Alternatively, Metacom. That is how the battle between Wampanoag Indians and white settlers came to be called King Philip's War. The English had noticed the "princely" attitudes of the Narragansetts when they dubbed Metacom "King Philip."

Along with their business mentality, Narragansetts could count to as high as 100,000, and apparently higher. Roger Williams, who traded with them daily, said that many or most could count that high. In that era, not even many Europeans could count to that level. He also noted that they were difficult to bargain with by being shrewd traders, *"… penurious to a degree. Will beat out all markets to save …"*

It was unnecessary for hunters to be familiar with higher mathematics but necessary for farmers and traders to do so. The Algonquian-speaking people used the same units as the Inuits, which evolved from counting the fingers and toes. A hand was five, a foot was five, and a person was 20. The count was started on the left hand passing to the right as the sun rises on the left and sets on the right in their folklore, according to Michael Closs in *Native American Mathematics*. However, not all Algonquins could calculate as well as the Narragansetts.

Narragansetts differed in another respect from other Indians. The chiefs were acquisitive and demonstrated their wealth by displaying their shell money (Wampum) on themselves and their spouses. They also had larger and more opulent residences than the commoners, a European feature, but not necessarily an Amerind feature of tribes.

Regarding Wampum, a Narragansett word, the Narragansetts not only manufactured money, but also regulated it over a wide area inclusive of other tribes. Not only was it standardized by width and length, in the earliest days of contact it actually had a rate of exchange. This might have been the result of a culture experienced in trade and commerce, which, of course, the Vinland Voyagers were. Vikings were avid traders, certainly the most widely traveled, and used coinage and other fine metals in commercial dealings. Did the Narragansetts take on some of their business savvy? Most "wampum" (peague) I have seen in museums is made of glass beads, also of white man's manufacture. I suspect that Narragansett manufacture and regulation of wampum from seashells was an indicator of remarkable sophistication, comparable in its way to ancient and original Chinese manufacture and regulation of paper money from mulberry bark.

The Narragansett mathematical high count is a highly unusual trait. Two hundred years later, it was found that Apaches could count to 10,000, and that was thought quite remarkable. There were very few in Europe who could do this. There, numerical ability was restricted to elites in the churches and universities.

This trait is not so much a question of latent abilities. For example, the Maya calculated to much higher figures for astronomical dating. With Narragansetts, they had to count items for manufacture of Wampumpeague and other things involved with the regulation of their people. It may be recalled that they participated in the 1670 census count.

Roger Williams was very impressed by this ability. He said of the Narragansetts,

> *Let it be considered whether traditions of ancient forefathers or nature hath taught them Europe's arithmeticke.*

Still another indicator that these Amerinds had European roots comes from Williams' book about Indians and their language. He wrote this remarkable passage: *"Mosk"* or *"auku'nawaw,"* the great Beare, or Charles Waine, which words Mosk, or Pauku'nawaw signifies a Beare, which is so much the more observable, because, in most languages that sign or constellation is called the "Beare." So unusual did this quite extraordinary anomaly strike Williams that he noted it in both his preface to the reader and later in the appropriate text.

It might be one thing that these Amerinds had a defined constellation that they identified as a bear. What is more noteworthy is that it was located in the same sector of the sky where Europeans described it. Moreover, when one looks at those stars, it is like a Rorschach card, which can be described as anything.

The English settlers, including Williams and his compatriots, referred to the constellation as "Charles Waine" and not as a bear, so the Amerinds did not learn it from them. *Ursa Major* (big bear) or Charles Waine is a large constellation in which the plowshare or the bucket-shaped Big Dipper of seven stars appears and the handle of the dipper forms the bear's tail. To find that the same pattern of stars as those described by Europeans, who derived it from the ancients, as a bear is amazing. Many other cultures have called the constellation by other names, but the Greeks and Romans had great respect for the powerful bear, making it almost into a god.

The translation of Ovid's *Metamorphosis* reads, "Those seven stars which take that name of the plowshare called vulgarly Charles Waine." Edmund Spenser's *Astrophel* elegy said "and Charles-Waine refus'd to be the shipman's guide." William Shakespeare's *Henry IV* Act 2, Scene 1 begins "Charles Waine is over the new chimney." The Amerinds must have gotten this from Europe!

In fact, that finding inspired Roger Williams to write this humorous ditty:

> *The very Indian boys can give*
> *To many stars their name,*
> *And know their course, and therein doe (do)*
> *Excel the English tame.*

Quite probably, Williams may have had some correspondence with Cambridge University concerning his Narragansett adventures. It is well within the realm of possibility that some Narragansetts may have visited the English campus with Williams. While we do not know if any Amerind friends accompanied Williams in his visits to his country of birth in 1643 and around 1651, England was no stranger to a number of Indians such as Pocahontas from Virginia and Squanto from Massachusetts. In fact, Williams' mention of Narragansett star names was likely determined by his university education.

Europeans, of course, had an established science of astronomy from "Araby." In fact, we know of Ptolemy's *Almagest* only through Arabic translations from Greek. From ancient times, those whose life and livelihood kept them outdoors for extended periods were close observers of the heavens. The naming of the stars and constellations may have been done by many cultures but it is exceptional that they would be the same across cultures in view of their indistinct outlines.

Numbers play a role in farming and husbandry, of course. Narragansetts were skilled farmers as were, indeed, a rather large belt of Amerinds west and south of their general district. Their skill at growing crops was superior to that of the incoming Europeans whose own skill was in animal husbandry—near absent in most of the New World. Europeans (including Vikings) had the advantage of having cattle, sheep, goats, pigs, and other animals.

Not many are aware of the advanced state of Native American agriculture in this district of America. Iroquois and Huron in particular were outstanding farmers and stored surplus grains in huge underground pits. The Plymouth settlers in their explorations of Cape Cod found the same situation, but with smaller pits. Most of these pits had been abandoned because of the drastic drop in population due to epidemics. The food recovered from these Amerind caches were actually the prime reason for the survival of the colony.

The Narragansett Amerinds, judging from contemporary illustrations, formed their fields in regular geometric patterns and tended them carefully. (The development of geometry no doubt arose because of farming and property requirements.) The women were responsible for tending these crops but were aided by the menfolk as well. The situation became a cause of disagreement between the Pequots and Connecticut colonists when the colonists' cattle and horses invaded the unfenced Indian fields. Many such instances were resolved when the colonists declared war on the Indians. This kind of advanced agriculture rarely existed in northern New England or in the South where the natives were more typically woodland hunter/gatherers.

Burial Differences

Archaeologists have found that Indian cemeteries on Conanicut Island and Grassy Island at Taunton River in Massachusetts contained cremations from the Archaic period through the 1500s. These preceded any contact with Vikings, who also had cremation ceremonies.

Narragansett burial practices are seen in a 1660 cemetery; site RI 1000, described elsewhere. Archaeologists found that "nearly all" individuals were buried in flexed (fetal) positions on right sides and with heads pointed toward the southwest. At least two of the cemeteries were laid out southwesterly. This indicates that they were not Christian, but did have a strong religious or social contract among themselves.

Nathaniel Philbrick in *Mayflower* described another burial ground uncovered in Warren, Rhode Island. It contained 42 graves of 17th century Narragansetts. They were also buried on their sides with their heads pointing to the southwest. Philbrick commented on page 190, "The artifacts collected from the graves provide fascinating evidence of the degree to which Western goods had become a part of the Indians' lives."

Philbrick noted that one of the graves had a richer assortment of objects. Included among the grave goods was a copper necklace. Edward Winslow had given Massasoit (father of Metacom or King Philip) a copper chain in 1621. This may have been his grave.

Roger Williams described how Narragansetts revered the southwest direction. Their legends held that corn had originated there and was transported to them by crows. The mildest winds seem to originate in that direction and they held that their souls would travel there after death. They had a word for soul (Cowwe'wonk) and a class of priests. They believed in a single omnipotent God who oversaw a number of lesser gods, which idea is very close to pagan Viking beliefs.

It might be noted that one of the Narragansett religious beliefs held that there were several indications of similarities with Christianity. One was a word "Pausuck naunt manit" which means "there is only one God," followed by several other more ambiguous words in Williams' book which may be repetitions of missionary teachings. He also remarked on a Narragansett legend of a God who walked on water as well as a destructive flood.

Europeans (Christian) usually were buried supine with arms at sides or on the abdomen or chest. At least two Pagan Scandinavian burials had bodies placed in the identical position as in RI 1000 and were oriented southwesterly. The grave goods of the two ancient Vikings indicate that they were warriors of some recognized social stature. In some aspects of Scandinavian medieval legends, the southwest direction seems to have had some significance.

The cemetery RI 1000 was of a neat and orderly layout and unique in its geometry, being long and narrow with burials arranged four abreast (with two or three isolated at remote places—perhaps unbelievers, Christian converts, or pariahs). Thus, its geometric regularity is not in imitation of the invaders, but was their own design. It is unknown how deep the original graves had been excavated, but the history of the plot and the fact that a bulldozer operator discovered it indicated that the graves were relatively deep—at least four feet.

There are records of burial elsewhere of a "queen" who was buried in an elaborate grave to a depth of 8 feet and records also indicate another "princess" who had been buried 12 feet deep! A curious observation might be surmised in that Viking ships were compartmented into "rooms," the spaces between the ribs in which normally 4 to 6 persons would reside at sea with two oars to the "room." The cemetery configuration recalls this peculiarity, which is something that might well be a cultural holdover. In fact, there are ship-shaped cairns, possibly burials, extant from olden times quite near or within Narragansett territory.

Amerinds elsewhere seem to have preferred the round in their councils and structures and cemeteries. Their fortifications were more commonly circular. Shallow excavations typified burials. Some burials were under mounds, either rounded or in animal or snake shapes called effigy mounds.

Elsewhere sometimes "burials" were in trees or scaffolds. It might be imagined that this was the result of the difficulties of digging in the absence of steel or iron shovels. Narragansetts in 1660 did have these utensils, but it remains a mystery explicable by anthropology alone as to origins of the culture-wide impulse to bury so deeply, so orderly and elaborately, and with such a wealth of grave goods.

European cemeteries are typically geometrically arranged to suit land adjacent to churches—square to rectangular. Ideal depth was six feet, but they were often shallower and seldom deeper. Burials in Scandinavia including Iceland seem often to have been, before Christianity, rather haphazard with individual graves outlined and even cairned in boat shaped form. The Christian cemetery at Brattahlid in Greenland had the bodies in supine position but also in a long and narrow cemetery configuration, while not yet in so orderly a geometry as RI 1000. Burials of powerful persons in pagan times were elaborate. A dead chieftain was often interred aboard ship with the whole ship buried along with possessions, slain servants, and horses.

Grave goods from RI 1000 and literary records show that copper for ornaments was predominant among Narragansetts, but to some degree Wampanoags as well. Free-found copper was common in the area and at least one copper mine of colonial times or earlier was at a place called Somers in Connecticut.

Records of an early explorer of the coast indicate that the natives of Martha's Vineyard told him that some of this copper was excavated from pits, which hints at knowledge of ore refining. Did the Vikings introduce their ore smelting to the Amerinds? Eastern Amerinds all used copper to some degree but not to as high a refinement or prevalence as Narragansetts.

The excavations in *L'Anse aux Meadows* in Newfoundland showed that Vikings had smelted bog iron ore to produce iron. There they had plenty of trees to produce the tremendous heat needed to melt the ore, although there was a shortage of trees in Greenland and Iceland.

The Narragansett custom of clearing land with bigger fires than other Indians suggests that they used fire easily. The worked copper sheets that they showed Verrazzano in 1524 suggest that they either smelted copper or traded with the Native Americans in Michigan who smelted copper after 1000 A.D., probably having learned it from the Vikings.

One grave exhumation of a "princess" showed her wearing leather-lined copper moccasins. A Wampanoag warrior was buried just a short distance from Narragansett Bay waters with so much copper that he was said to have been entirely encased in it (not true, but still sufficient to be termed plentiful).

Archives from Jamestown, Virginia, founded in 1607, indicate copper as being a major commodity in trade with the Indians of that area. If the Indians desired it so much at that distance of 300 miles, this demonstrates that smelting must have been a rarity among the natives.

Narragansetts often burned the possessions of individuals who had died of disease or were victims of sickness. Grave goods show that this was not universally done but only when severe contagious disease was suspected. This seems to infer some understanding of germ theory or at least transmission of disease. Roger Williams described it and felt that it may have contributed to their survival.

> *This (burning) the other Indians about us approve of as good, and with their sachems would appoint the like: and because the plague hath not reigned at Nanohigganset (Narragansett) as at other places about them, they attribute to this custom there used.*

Many other Indian interments elsewhere had evidence of grave goods, mostly of personal possessions. Others did not seem to have any awareness of disease transmission as evidenced by their ready acceptance of contaminated blankets and implements from European colonists. Apaches burned all possessions of the dead and went to the extreme of consciously "forgetting" everything about the person, including the name! It was a very decisive taboo, causing some embarrassment to anyone not knowing that the person had died.

The Wampanoags believed this very strongly. When King Philip (Metacom) heard that an Indian with the white name John Gibbs on Nantucket Island had mentioned his father's name (Massasoit), he sailed there to kill him. The path he ran down to approach his victim is still called "King Philip's Run." Peter Folger, the grandfather of Benjamin Franklin, saved Gibbs.

Europeans were well aware of disease transmissions and were sometimes in the habit of slaying survivors of shipwrecks in dread of contagion. Shipping from infested ports was constrained, quarantined, or forced away by Europeans.

Roger Williams described the shunning of ill Narragansetts:

> *In cases of sickness, their misery appears, for they have not a raisin, nor a currant, nor fruit, nor spice, nor any comfort more than their corn and water, wanting all means of recovery or present refreshing. I have been constrained beyond my power to refresh them, and I believe to save many from death. The visit of friends was all their refreshment under such conditions; and their visits did not occur when a disease was thought infectious. Then all forsake them and fly. I have often seen a poor house left alone in the wild woods, all being fled.*

Verrazzano attended a funeral of Narragansetts in 1524 where much grief was demonstrated and remarked that their dirges sounded "Sicilian." They were especially bereft if the victim was a child.

Amerinds elsewhere believed in the efficacy of varied natural herbs and balms, so much so that the term "Indian Snake Oil" entered the vernacular. Many of these treatments were impotent, of course, but obviously many were effective since most medicines until recently had natural bases. This is a most striking contrast in cultural belief within such a small territory.

By good fortune, the Arab chronicler Ahmed ibn Fadlan in 922 A.D. remarked and described the identical attitude as Narragansetts among a group of Vikings along the Volga.

> *When one of them falls ill, they erect a tent away from them and cast him into it, giving him some bread and water. They do not come near him or speak to him; indeed they have no contact with him for the duration of his illness, especially if he is socially inferior or is a slave. If he recovers and gets back to his feet, he rejoins them. If he dies, they bury him, though if he was a slave they leave him there as food for the dogs and the birds ...*
>
> *Their principal chieftain, a man of the name of Wyglif, had fallen ill, and was set up in a sick-tent at a distance from the camp, with bread and water. No one approached or spoke to him, or visited him the whole time. No slaves nurtured him, for the Northmen believe that a man must recover from any sickness according to his own strength. Many among them believed that Wyglif would never return to join them in the camp, but instead would die.*

His description of the funeral of the leader is well known, but the foregoing events and disease which killed him less so. Viking belief was known to be stoic in the face of death from any cause. Many modern Scandinavians in my experience seem to share this attitude, with the sole exception of hearty doses of cod liver oil for myriad ailments.

The Vinland sagas give a rather long and comprehensive tale of the death of Thorstein Erickson at the Western Settlement with his wife, Gudrid, in attendance. No succor was attempted while the unfortunate Thorstein died of some pestilence. In fact, he seems to have been "helped along" by the surviving Thorstein the Black, with no attempted hindrance by the observing Gudrid.

Sexual and Sanction Similarities

The Narragansetts meted out judgments and justice on themselves and outsiders in somewhat unexpected and different ways than other Amerinds. Their social order and the elevated role of women were quite different, perhaps stemming from the Vikings.

In 1660, there were but four English settlements in the colony, the two on Aquidneck, another in Warwick, and a fourth at Providence, which had been negotiated by Roger Williams. Colonial settlers were welcomed in these places. At these locations, the settlers formed buffer colonies between Amerind tribes across the bay. In 1660, there could not have been more than a hundred or so, if that many, Caucasians within Narragansett territory.

Narragansetts were so cordial to the settlers that until King Philip's War in 1676, colonists passed freely through and around Narragansett territory without hindrance. The Narragansetts were in the process of integration and often wore English clothing. To the west were several other groups called Nipmucs, Niantics, and Naticks who also spoke a dialect of Algonquin. These offshoots of the tribe were also amicable to colonists.

Rhode Island was open by invitation to exiles and escapees from Massachusetts and Connecticut from about 1630 to 1676. Grave goods buried with Amerinds during that period were examined at the RI 1000 site, and indicate both friendly and intimate relations to Rhode Islanders. In fact they show considerable contact with distant European colonies as far west as the Hudson River and as far north as Lake Champlain.

There is common ground in sociological practices between the Narragansetts and the natives of Martha's Vineyard, particularly that group called "Gay Head Indians." For example, the hierarchal social structure of land use was by social groupings. Commoners had to pay if they pursued game into adjacent lands of a noble or chieftain not their own. This is something unusual in normal Indian behavior where land is not private property. It resembled land use in Europe, which was rigidly regulated by property ownership.

Along the same line is the Narragansett custom of commoners being executed by leaders or agents of leaders. In Europe, executions were common and popular diversions. Europeans seem to have been inured and fatalistic to executions. Narragansetts took heads as trophies and, according to Roger Williams, did so with great dexterity and "... even with a sorrye knife."

Until quite recent times, the taking of heads and displaying them trophy-like was common among Europeans; after all, the guillotine was still in active use as late as the 1930s. Most medieval cities and towns had places of execution and portions of walls and pikes thereon for the display of heads and even bodies.

This was common practice among the English and Dutch colonials and, in fact, was a contributing irritation to the Indians, which led to King Philip's War. A popular female Wampanoag chieftain's severed head was on display for some time at Plymouth, and her relatives were said to have demonstrated "much rue." It would seem that the habit might have originated in sword-wielding societies. Despite the apparent barbarity of the practice, it aligns Narragansetts more closely with Europeans than with Amerinds, since it seems rare and unremarked in other North American tribes.

A common myth is that scalping originated with American Indians but that may not be the case. Many historians believe that scalping was introduced into the Indian culture by the colonists themselves for the purpose of paying bounties. Bounties for Indian scalps are recorded in New England but it is not clear who originated the practice.

Little observed is that some cultural items we automatically presume to be Indian are, in fact, of European origin. Indian tepees of plains Indians as we see them are of canvas—"white man's commerce"—with an improved covering replacing heavy and unwieldy skins of the same basic design. In addition, the Indian tomahawk! Not Indian at all. It was a useful weapon in medieval European wars and the tomahawk is apparently a reduced design of Danish battle-axes. Colonial blacksmiths could manufacture at least crude forms of them, but the Indian came up with the form combining ax with tobacco pipe.

The status of Narragansett women was high. There was a high moral standard among them and both Roger Williams and Giovanni da Verrazzano recorded this. Women could become chieftains in their own right. In unconscious humor, Williams recorded that women tended most of the fields (except tobacco) and that the men aided them "... even when they did not have to." A captive of a Narragansett Sachem during King Philip's war, Mrs. Rowlandson said much the same and stated that in her several months captivity she had been deprived, but never molested, even by her captor.

We can assume that when new arrivals intermarried with natives, they also took native women for wives and servants. This was the custom of explorers in the past and a characteristic of Vikings in particular. The enthralling books of the late Peter Fruechen demonstrate just how easy and natural this was to do.

In a similar manner, Narragansetts took women and children captive as they did Mary Rowlandson, but this was rare in that area and those early days. Interestingly, a century later, Benjamin Franklin remarked on the ease of both male and female whites "going Indian." He added that of the "many" that had migrated to the supposedly more primitive culture, none ever voluntarily returned. These people in New England became colloquially known as "Swamp Yankees," from whom this writer is descended.

Quaiapen (the old Queen) was a daughter of Ninigret. She married Canonicus' son, participated in political decisions, and assumed leadership responsibilities in King Philip's War until she died. Her head was severed from her body by John Talcott's troops and mounted on a pole where Indian captives were being held. Their laments and mourning stirred Indians into an attack on colonials that culminated in the Great Swamp War.

Another daughter of Ninigret was Weunquesh, who succeeded Ninigret ruling until her young brother, Ninigret II, came of age. The male and female sachems protected their followers from internal and external dangers, allocated land, provided for the poor, and administered justice to wrongdoers, according to the *Handbook of North American Indians*, Volume 15, edited by Bruce G. Trigger.

The fact that Narragansetts had not only some, but many female chiefs is also highly significant to us. This trait was also common among the Vikings/Northmen and could well be a holdover from past invaders who intermarried with Native Americans. Mrs. Rowlandson's captor seems to have treated her with considerable respect. While she was starved and her infant died in her early days of captivity at some distance from the Newport area, eventually her position improved and she was never molested during those 11 weeks.

Elsewhere, there seems to have been varied attitudes by Amerinds toward women. It was highly unusual for Indians elsewhere to have female chieftains. Apache behavior was notable in attention to women's prerogatives, even conducting feminine puberty ceremonies despite being closely pursued by enemies. Plains Indian women owned the canvas or skin-covered tipis and their husbands left their own and entered the wife's family. However, I have not discovered any female chieftains among other tribes. As far as I have been able to determine, most Indian cultures were "meritocracies" with chieftains followed either because of their skill in war or hunting, or sagacity in reasoning at councils. For an individual, following a particular chieftain was entirely voluntary.

European attitudes also varied but usually women occupied a low social role except for Viking women, who could become leaders in their own right. This social trait is most unusual elsewhere in Europe. As servants, European women often, if not generally, led a hard life. American colonial women especially had difficult times and high death rates, what with hard labor and frequent childbirth. A great number of colonial males went through a series of three and more wives. Women pioneers only occasionally lived to old age in the early days. However, Scandinavian attitudes in Europe were different. The role of women was notably favorable as can be seen by the life of Gudrid Thorbjornsdottir. This cultural attitude compares closely with Narragansett behavior.

Roger Williams told an interesting description of how a father reprimanded his son. Roger's Narragansett friend asked his nine-year-old son to get a drink of water for the visitor. As with many other cultures in close contact with nature, Narragansett parents seldom disciplined their children. In this case, the boy snappily retorted to his father that he should do it himself! It appeared that this would be the outcome but Williams, interceding, advised the father to respect his own position and establish his authority. This the father did with the result that the boy picked up a stick to defend himself, and the situation escalated with the two of them exchanging blows. The father won and remarked to Williams that it did seem to be the better way, the boy did get the water, but Williams seemed to imply that the lesson was not long lasting.

Narragansetts were well armed (even with muskets in 1635) and carried their arms habitually. They were known to have been militarily capable, yet were not overly belligerent in their social affairs even with outsiders. Their propensity for welcoming strangers was remarked upon by Williams and Verrazzano, as well as the factor of 50 years peaceful, indeed benevolent, cohabitation with the nascent Caucasian Rhode Islanders.

Roger Williams accompanied a party of more than 200 on a trip to Hartford, Connecticut, through enemy territory. He said:

> *I have upon occasion traveled many a score, nay, many the hundredth mile amongst them, without need for a stick or staff, for any appearance of danger among them; yet it is a rule among them that it is not good for a man to travel without a weapon nor alone ... I have known them to sleep outdoors that their guests (invited and otherwise) might be sheltered.*

This philosophy among other Amerinds was quite variable depending upon the status of a group and its association with neighbors. Indian warfare, however, was never outright genocidal, although treatment of prisoners of war was often severe.

Strange to say, this attitude is in common with archaic Vikings. Medieval writers often remarked on the ease and safety of travel in Scandinavia in olden days. One remarked that even the poorest average person would share his last crust of bread with visitors. Yet Vikings also were habitually armed to the teeth, never without a knife and usually with sword and shield abroad. An insight into Viking custom is preserved in Scots traditional dress with cape over the shoulder and with the dirk in the stocking.

A Lakota (Sioux) woman named Madonna Swan contributed information concerning Native American customs and how they differed from Narragansett social practices. For example, the Narragansetts were a strictly hierarchical culture and did not have the practice called "Potlatch" which Swan described. Let me describe the purpose of the "Potlatch."

I have pointed out before that generally the Amerind before 1492 was extremely individualistic and democratic. Most tribes would not take action unless there was general agreement among them. War chiefs could be censured and deposed if losses were excessive. Chiefs could be deposed or abandoned if the tribe met misfortune, calamity, or poor hunting. The result of this was that if chiefs wanted to maintain their position, they had to make constant efforts at appeasement of their followers.

One way to appease was to demonstrate generosity by giving away goods and food. This was so prevalent that early explorers could often identify the chief by looking for the plainest and poorest man in the village, and usually this was a male with few exceptions. The practice was only recorded late in history in the

Northwest coastal tribes of Washington and Oregon where the occurrence was termed "Potlatch."

Madonna Swan described the practice in several places in her narrative and, indeed, did it once herself when she had become a somewhat influential person on the Cheyenne reservation in South Dakota. She termed it "give-aways" and used it to maintain position. These descriptions suggested that this social phenomenon was common among Amerinds across the Americas. Certainly, Narragansett chiefs who had the power over life and death did not believe this a necessary practice.

WATER AND LAND USE DIFFERENCES

According to archaeological records, the Pettaquamscutt River Basin was unoccupied for eons until the sudden appearance of an abrupt change in living style about 1,000 years ago. Archaeologists described it as violating the usual slow gradual pace of evolution. Almost suddenly, there were dwellings and campsites up and down riverbanks and salt-water shores. This was unique in those times because it required a people who had a certain degree of nautical skill and military prowess. They suspected a large enough incursion of foreigners to cause abrupt changes in ancient ways.

Because of so much archaeological research and early colonial writings in this area, it now appears that we know more about this Narragansett group than almost any other from 1000 to 1700 A.D. We can determine the date that Narragansetts commenced their distinctive development. This comes about from examination of their earliest chroniclers and the modern assessment of their ancestral locale—the Pettaquamscutt River.

Verrazzano found a populous, healthy, and advanced culture. It was united by language, law and social status all firmly established in their territory. He could learn of no recollection or legends of their origins. They habitually dwelt at sizable oceanside estates of perhaps 50 acres kept clear of prolific growth by systematic burning programs. So large were the settlements that Verrazzano said an "army could maneuver" in them. While they no longer exist, their numbers and locations can be identified by their still extant Indian names.

A century later, Roger Williams set up a trading post within the Narragansett domain—apparently the second of only two in a 15 by 20 mile territory. The other post was Dutch. Williams found that they were nautically adept, confirming Verrazzano's descriptions of their finely crafted vessels and adding that their nautical territory extended all along the New England coastline and including Block, Martha's Vineyard, Nantucket, and Long Island.

Their nautical presence was felt all along the coast from Nantucket to at least the Connecticut River and included Block Island and Long Island. The natives of Martha's Vineyard ("Nope" in their day, later "Capowock"—our "Straumney") shared many similarities with them. Besides occupying Block Island, they also dominated several islands of Narragansett Bay, owning the island Conanicut (Jamestown) and almost certainly sharing Aquidneck (the Rhode Island Newport area).

By 1635, their population was some 15,000 to 30,000 persons. Their territory was west of the present township of Narragansett by some 15 miles; and from Block Island Sound north some 20 miles, making a territory of rather limited area and dense population. They were dominant not only in lower Narragansett Bay, but also for many miles inland and seaward. They were greatly respected by other tribes and the English, Dutch, and French colonists.

There was minimal enmity with Wampanoags who occupied the east side of the bay. Despite this, traffic was regular between opposite shores and intermarriage frequently occurred, similar to the Capulets and the Montagues in *Romeo and Juliet*. Indeed, the division of the groups on opposite sides would not appear valid. Those residing about the bay were essentially all the same people, differing only in some cultural practices and linguistic colloquialisms, idioms, and dialects.

It happens that Aquidneck Island has a very narrow strait at its northern end. It is likely that this large island, upon which Newport now exists, was territory held in common. This was fortunate for some colonials who felt uncomfortable in the new settlements east such as repressive Plymouth and Salem colonies.

Two of the original settlements of the present state of Rhode Island originated on that island—Newport to the south, and Portsmouth to the north end, near the straits. The island is of the size and general shape of Manhattan.

Massasoit, befriender of the Plymouth Colonists, actually was a resident of the shores of Narragansett Bay. While technically a Wampanoag, he was within the same gene pool as Narragansetts. He had a number of features about him more Narragansett than Wampanoag, including the relative immunity to diseases and possibly even a beard. When he first arrived at Plymouth, he had an entourage of some 60 persons (probably all males and warriors) and had traversed 45 miles of lands essentially decimated by disease without being affected.

He (his other Indian name was Osamequin) and his sons were born on Narragansett Bay at a place called Chacapocassett near Warren, Rhode Island. He resided later at Somerset, a bit west of Bristol, Rhode Island. At the time of the Plymouth contact, he seems to have been identified as "Pockonokick Sagamore" at what is now Dighton, Massachusetts. (Dighton is where a famous carved stone is preserved, which some say depicts European runes.)

Massasoit seems to have enjoyed considerable power over a wide territory, having sold the city of Pawtucket, the Island called Prudence, to Roger Williams. He also sold the entire island of Aquidneck to William Coddington. In fact, when Aquidneck was sold, some of the same Narragansett chieftains that had earlier sold Providence to Roger Williams signed the deed.

Narragansett vessels were built with "great artifice" according to Verrazzano and were large enough to hold 24 persons. Rowing or paddling was the more common means of locomotion but Williams says that sometimes makeshift sails were used. Verrazzano says the crew rowed or paddled in synchronal mode "without using their backs," making much use of their muscular arms. Williams was impressed with their seamanship and says they traveled habitually well out of sight of land and to great distances.

Their influence extended to at least Nantucket to the east and Connecticut River to the west and as far out as Long Island. They and other Wampanoags were valuable to early Nantucket settlers by showing them a shortcut through the long arm of Cape Cod that could be accomplished only in small narrow boats.

Their seamanship was noted when one Narragansett chieftain conducted a raid on Long Island with strikingly similar overtones to the commonly described Viking raids in Europe.

The Northwest Pacific Coast Indians shared this nautical ability. It could well be something that developed independently because of residing by a huge, sheltered waterway. Yet, the ability and tendency seems unusual among those we have always viewed as eastern forest dwellers in general and Algonquins in particular.

Viking seamanship needs no further elaboration here, yet it must be remarked that the propensity for the sea becomes ingrained in populations and is something that would be expected to be transmitted to descendants. The famous Viking "longships" were a development of a dugout form of craft, which the Narragansetts also used. The Vikings gradually added freeboards for use in rougher waters, and eventually built them all this way. Indeed, the presence of a top strake (highest side plank), of another color or type of wood is symptomatic of Vikings in particular. It demonstrated the difference between a fjord "galley" with sails and a sea-going vessel.

Narragansetts practiced systematic clearing by fire of large areas of forest when extra space was needed. This implies a close social organization and a high degree of scientific application. At contact, the forest growth must have been enormous, with numerous great trees of many species, whereas today the "second" growth is still moderately high but extremely thick, predominantly with maple and pine, frequently interspersed by oak. It is not an easy task to accomplish in a dense forest and requires attention to times and conditions. For example, the forest must not be too dry.

Apparently, these fires were done twice during the year, one season of which is still known in New England as "Indian Summer." This period is in October/November when temperatures rise to summer heights. The air becomes "smoky" yet at once both invigorating and enervating. In those days, it was smokier yet from the extensive fires.

They were otherwise quite skilled in using fire as Roger Williams informed that they used fire to hollow out their finely crafted vessels. In one case around 1660, a split-log sarcophagus with metallic hinges had been hollowed out by fire. Williams said that individuals would take themselves off into the forest and construct a vessel in ten or twelve days. The preferred wood was Chestnut.

Amerinds elsewhere may have used fire to this extent but I have been unable to discover other instances of it. It seems peculiar to Algonquins and New England in particular. It requires a measure of high advancement, strict social order, and must have originated at a time when iron tools for clearing trees were unavailable. Stone axes would not be efficient enough for extensive clearing of this sort.

I am unaware of extensive use of fire to this degree in Europe. In medieval times, much of the forest had already been cut for heating, cooking, and manufacture. Perhaps fire was more feared than otherwise. Clearing land by fire among the Narragansetts thus does not appear as a holdover but something original resulting from their own strictly organized society.

However, it must be remarked that the Vikings who traveled to Vinland were mostly born and raised in Greenland and Iceland where tree cover was nil or very sparse. It is unclear why Narragansetts used fire for clearing more than other Amerinds. Could this be a reason why Narragansett descendants seemed to have been less comfortable under the forest canopy when other eastern groups favored forest cover?

Clearings around residences and villages have defensive advantages as well as protection from fire. The English colonists were great hewers of trees, considered them "weeds," and felt that the forest had been made to be chopped down. They always cleared the environs of their home sites, partly from military considerations but also from fear of fire. The Romans, partly to deprive the natives of cover, had cleared England. Later Britons consumed great quantities of forest re-growth for heat and to manufacture glass.

Narragansett dwelling styles held remarkable similarities to medieval Viking styles. Narragansett homes were structured from bent saplings covered with woven mats, about 15 feet square, transportable, and moved quickly to escape insects or to be near fields and so forth. Only when Iceland and Greenland used up all their trees did they turn to sod homes or rock buildings. Their preference had always been wood.

Many Narragansetts occupied these small dwellings. Verrazzano said as many as 28, and Williams said he often saw two families residing in one dwelling "lovingly." Williams even described the names Indians gave to tame beasts and birds, which they kept about their houses. Of course, close quarters causes higher disease transmission. Their remarkably high survival rate indicates considerable immunity to disease.

The ruling class lived with extended families in an entirely different type of structure, a larger type of house called, as in Viking society, a "long-house." There were numerous "villages" interspersed all throughout Narragansett territory. Viking lifestyle was notably similar. The houses were more advanced and permanent but excavated Viking medieval houses in northern Europe were of wattle and daub construction. They, too, were so small that they must have been crowded.

Most Amerinds elsewhere did not practice the centralized type of dwelling as a long house and usually the norm was one family to a structure in a hogan, wickiup, tepee, or wigwam. "Wigwam" is another Narragansett word. Northwest tribes lived in plank houses; Southwest Indians sometimes in elaborate earthen enclaves or, in Mexico, even frescoed stone structures. Some primitive tribes simply lived in lean-tos or some haphazard arrangement of branches or brush.

In these hunter/gatherer societies, the population often moved as they exhausted the flora and fauna of their area so they did not set down the kind of permanent structures used by farmers.

European agriculture was advanced but with numerous small land holdings or fiefs. These were usually under control of higher social classes to whom the farmer was in debt one way or another. It was similar to Amerind practices but, as mentioned earlier, animal husbandry was more important in the Old World lifestyle. This matters to our study because the domestication of cattle (bovine or other ruminants) has great bearing on the development of tuberculosis.

Narragansetts were enthusiastic consumers of fish and shellfish prevalent around Narragansett Bay. This might be expected while comparing it with European medieval reluctance to eat shellfish even, apparently, near seaside residence. To Vikings, of course, fish was a staple food. To the Inuits, it was their main food source.

The trait seems unusual in other Amerind societies, which seldom lived in a seaside mode. There is the remarkable taboo among the Apaches to avoid eating anything that lived under water.

Although coal is not mentioned in the sagas, Helge Ingstad (who found L'Anse aux Meadows in Newfoundland) described how Rhode Island's type of coal was found in Greenland. He wrote in *Westward to Vinland*:

> *In the Western Settlement in Greenland, there was once a farm ... which around the year 1000 probably belonged to the Vinland voyager Thorfinn Karlseffni and his wife Gudrid ... One of the most curious finds was a lump of coal, which was dug up from the depths of the main building ... The lump was anthracite coal ... this type of coal does not exist in Greenland ... there are only two deposits of anthracite coal along the east coast of North America, and they are both to be found in Rhode Island.*

Anthracite coal has the highest amount of carbon in coal and is easy to light, burning a short blue flame. Since the Vikings were already doing smelting by the time they were at L'Anse aux Meadows, they might have been familiar with that form of coal. It was not used in the Americas until the 1700s.

In 1295, Marco Polo noted Chinese use of coal as if it were new to Europeans, yet this Greenland/Rhode Island coal pre-dates that information. Perhaps some enterprising scholar might investigate the possibility that coal use was introduced to Europe from the New World. Such an idea is far from inconceivable.

Coal is ordinarily difficult to ignite, looks like stone, and an unwary explorer who used some black stone as a fireplace might easily discover its thermal properties. Anthracite coal has a particular luster that attracts the eye. However, unless it is brought into close proximity with a wood fire, it simply looks like black shiny stone.

Paul Chapman also reported about the finding of anthracite coal in Narragansett Bay in the article "Norumbega: A Norse Colony in Rhode Island" in 1994.

Chapman also stated therein that he believes the Narragansett Indians are descendants of the Norse.

That finding and other evidence led Marshall Smelser to state in his 1967 article "Medieval Europeans in America: Latest Findings:"

> *In the new world they [Norsemen] seemed to have coasted Hudson Bay, Hudson Strait, Baffin Bay, Davis Strait, and the Atlantic as far south as Rhode Island in leaky ships up to 166 feet long and which could not sail against the wind.*

Let us look at the issue of grapes (vin) in Vinland. The issue of grapes in Vinland has come under much discussion since the discovery of the Norse site at L'Anse aux Meadows. The discoverer, Dr. Helge Ingstad, initiated this question in furtherance of proving that this site actually must have been Leifsbudir. He proffered alternative linguistic meanings of "vin." However, the section in the sagas that initially identified grape vines and grapes is one of the clearest in the sagas—almost as if the chroniclers wanted to make doubly sure of their description. Grapes were also mentioned by Thorfinn Karlseffni, and also by Thorhall the Hunter, and in such a way as to confirm the first remark. Later mention by medieval historians was also explicit on the matter.

The importance, of course, stems from their earliest days of conversion to Christianity when Catholic doctrine insisted that wine, and grape wine alone, must be the blood celebrated in masses. Wine being such a rare and expensive commodity in the Northlands, a Viking discovery of their own sources was phenomenal good fortune—as good as, if not better than, gold. I will comment here on the existence of grapes in Rhode Island but with a comment first about another site.

Martha's Vineyard, with Chappaquiddick, Noman's Land, and the Elizabeth Islands comprise Dukes County, which was incorporated November 1, 1668. The first governor, Thomas Mayhew, who was hoping thereby to gain royal favor, named the county for the Duke of York. There are six towns on Martha's Vineyard. Edgartown on the east was named for Edgar, son of James II, who bore the title of Duke of Cambridge. Oak Bluffs on the northeast was named for its location and oak trees. Tisbury was named for the Mayhew Parish in England. Chilmark was so named for the English Parish of Governor Mayhew's wife. Gay Head on the west was named for its wonderful cliffs of different colored clay. Martha's Vineyard was named for Mayhew's mother, Martha, and it was noted for its grapes. This is why Martha's Vineyard is sometimes suggested as the possible site of Vinland.

Do grapes grow in Rhode Island? Yes, they do, and have for hundreds of years. Indeed, they are exceedingly plentiful at Pettaquamscutt, some years blanketing huge areas of the forest floor as well as climbing trees. I have seen one hardy specimen attempting to thrive even along a salt marsh shore. Some are feral in these days, but the native grapes are easily identified by being smaller and not so sweet. The sagas say "... they filled the after-boat with them," thus signifying a quantity easily achievable at Pettaquamscutt. Since they have at least some sugar, it surely must be possible to extract a wine from them even if not the refined taste of modern wines. In older days, an Epicurean taste would be a luxury reserved only for nobility and likely for them only sometimes.

Colonial settlers tried to make wine from the wild grapes, but gave up when they could not make it taste like the fine wines of Europe. If the country could grow great quantities of wild grapes, then why couldn't they make good wine, they wondered? The answer was that the New World had Phylloxera vastatrix (a root louse), and a lack of hardiness against the cold winters. These two reasons held back wine production in the New World for several years until they were able to rid vineyards of the tiny insect.

Today there are 22 wineries in New England and over 30 vineyards. There are four active wineries in Rhode Island: Sakonnet Vineyards in Little Compton, Diamond Hill Vineyard in Cumberland, Vinland Wine Cellars in Middletown and Prudence Island Vineyards on Prudence Island in Narragansett Bay. Besides producing consistently high quality wines, the wineries are all in very green and beautiful areas of the state, just beyond the traffic but close enough for easy transportation.

Aquidneck Island in Narragansett Bay is the official island in Rhode Island's name. Fifteen miles long and five miles wide, it supports three communities: Portsmouth to the north; Middletown, the home of Vinland Wine Cellars; and Newport, at the extreme southern tip. A visitor can approach Vinland Wine Cellars by way of Route 136 South, with views of the river, prosperous farms, stone walls, antique shops, and the Portsmouth Historical Society. Although it takes its name from an ancient Norse settlement, Vinland is the newest winery in New England, just opened to the public in 2006.

John Marks, writing in the Internet site Stereophile #28, *The Fifth Element*, Feb. 2005, had this to say about wine in Rhode Island:

> *The Vikings called the North American area they explored Vinland because they found grapes growing there ... There is also geological evidence that the Vikings got at least as far south as Narragansett Bay in Rhode Island ... Narragansett Bay,*

and the north and south shores of Long Island Sound, are several degrees of latitude south of the Burgundy wine region of France. The coastal regions of Rhode Island, Connecticut, and Southeastern Massachusetts all qualify as good candidates for the historical Vinland.

The issue of "self-sown wheat" has been important to those who try to pinpoint the location of Vinland. The sagas describe how the slave runners brought back not only grapes but also self-sown wheat. That is also called "volunteer wheat" and the generic name is *Triticum aestivum*. Nowadays, most farmers know that this particular wheat carries a virus and they try to eradicate it. However, that virus would have been unknown and the sheaf of wheat would have represented to Vikings a source unavailable to them in their cold clime.

Preserving food and drink always was a problem back then. Norwegian culture today still offers the ancient food called "Lutefisk" which is codfish steeped in lye (wood ashes). It is served at Scandinavian and Northern German banquets as a symbolic food and, indeed, some people actually like it. This writer confesses that he prefers it to fresh fish. I am not sure that Narragansetts preserved fish or used lye. Preservatives for long sea voyages were a necessity for food.

Conclusions about Narragansett Differences

The unique features mentioned above call attention to various differences between Narragansetts and other Amerinds and similarities to Vikings. I believe that some of the most compelling information about why Rhode Island was the site of Viking landings and interbreeding has to do with the language similarities, the land use differences, the finding of anthracite coal in Rhode Island which was carried home by Vikings, and physical differences.

I will add a little more about physical differences based on personal observation. I have interviewed a full-blooded Narragansett woman (D.L.) who has fine black hair but none of the facial characteristics associated with Indians, although her brother does. Her three children are blond, as is her husband. One would never suspect by looking at her that she is a full-blooded Narragansett and I believe that she may be a result of the Norse incursion. She has never bothered to study that subject and was only brought to my attention by her brother.

Not only are there descendants from Indian Viking interbreeding, but there are descendants from Vikings who can trace their ancestry to people in the sagas. I heard from a Peggy Marie (Lee) Woods who lived in Broken Arrow, Oklahoma, in the 1990s. Peggy Woods could trace her heritage through her great-grandparents, Einar and Gudrid Eiriksson who came to America from Iceland in 1880. I will use the spelling she uses for her ancestors. She has records going back to

Gudrid Thorbjornsdottir (Leif Erickson's sister-in-law) who married Thorfinn Karlseffni Thordarsson. She comes down the line not from Snorri Karlseffnisson who was born in Vinland but from his brother, Bjorn Thorfinnsson, who was apparently born after Gudrid left Vinland.

To this day, some descendants live on and tell their tale. One example of descendants comes from *A Historical Sketch of the Carr Family from 1450 to 1926* by W. L. Watson, which states the following introductory information:

> *Chief Canonicus was my 13th great-grandfather ... In the book Bearse-Bears-Barss Family, Genealogy of Augustine Bearse and Princess Mary Hyanno by Franklin Bearse, it tells of the Vikings coming to the Wampanoag area about 1001–1016 ... They were fierce, red haired, pale-faced men who came, to what is now Massachusetts, mixed their blood with the Wampanoag Indian ... "Wampanoag" means "White Indian." Mary Hyanno (born 1625) was of light complexion and had flaming red hair. These stories were written from the legends passed on from one generation to another in the Wampanoag tribes.*

I will add a little more information about legends about Norse-Indian contact. Legends of the Wampanoags on Martha's Vineyard Island (the possible site of Straumney) tell of early visits by a very different people. Matthew Mayhew, grandson of the Governor of Martha's Vineyard, Thomas Mayhew lived from 1648 to 1710. He preached and converted natives on the island and learned their history in the process. He describes them in his book published in London 1694–1695: *The Conquests and Triumphs of Grade: Being a Brief Narrative of the Success Which the Gospel hath had Among the Indians of Martha's Vineyard.* One passage of that book explains that some on the island had a different heritage with lingering effects:

> *Although the people retained nothing of record nor use of letters, yet there lived among them many families, who, although the time of their forefathers first inhabiting among them was beyond the memory of man, yet were known to be strangers or foreigners, who were not privileged with common rights, but in some measure subject to the yeomanry, but were not dignified, in attending the Prince, in hunting or like exercise, unless called to particular favor.*

One of several stories that Martha's Vineyard was the site of a Viking landing came from Charles Edward Banks' 1911 multi-volume work entitled *The History of Martha's Vineyard*. This passage describes the legend of white men with red beards coming to the island. The grandmother of a Wampanoag given the English name of Thomas Cooper passed it down. In 1792, sixty-year-old Cooper told Benjamin Basset, a selectman from the Vineyard town of Chilmark, of the oral tradition:

> The alien came in great canoes, larger than any they had seen before, and they were filled with Wauta-conua-og; men with coats of another color from the east. They brought with them strange weapons, fashioned out of a new material, had red beards; and knew not Michabo or Cheepie [native gods]. It seemed they were the expected people who were to come out of the rising sun from the dawn, white ones like Michabo, and inhabit their hunting and fishing groups.

Another reason that Vikings might have stayed in North America rather than returning home has to do with climate. During the 1200s, Greenland experienced horrifying frigid climate changes that made it less desirable as a home. Since I moved from Rhode Island to Arizona to get warm, I would not be surprised if some Vikings opted for the warmer climate along coastal America rather than return to Greenland.

Voyages to Markland (Labrador), probably to cut timber for ships and fuel, continued until at least 1347, when a ship bound from Markland to Greenland is recorded to have been driven off to Iceland. It was described as a "small, Greenland ship with 17 men aboard, who had lost their anchor" and thus required help to beach or moor, and was laden with timber from Markland. Neither Iceland nor Greenland had sufficient wood for shipbuilding and for the many other uses of wood. Therefore, voyages between Greenland and America took place for at least 347 years. That is almost the life span of the United States; i.e., from 1653 A.D. to 2000 A.D.

One of the most important differences between Narragansetts and other Amerinds is their genetic resistance to tuberculosis. They may have had resistance to other diseases as well. Nathaniel Philbrick said on page 49 in *Mayflower:*

> Before the plague, that had numbered about twelve thousand, enabling Massasoit to muster three thousand fighting men. After three years of disease, his force had been reduced to a few hundred warriors. Making it even worse, from Massasoit's perspective, was that the plague had not affected the Pokanokets' neighboring enemies, the Narragansetts, who controlled the western portion of the bay and numbered about twenty thousand, with five thousand fighting men.

This can only come about by contact with those who had been exposed to the disease in Europe.

Sidney Smith Rider (a pre-eminent Rhode Island historian, 1833–1917) wrote that the Narragansett tribe was one of the two "greatest" tribes upon the North American continent. The Narragansetts were the largest of the New England tribes and were respected for their ability to field 5000 warriors when other tribes could not. They were prosperous, numerous and healthy. It was noted that not only did they survive the European diseases that were decimating other tribes but Williams and colonial Connecticut Governor Winthrop both stated, "The plague did not seem to be among them."

By 1660, American aborigines in New England were suffering the effects of personal, psychological, and social debilitation from diseases and many groups were wiped out. In the census of the Narragansetts around 1670, they still had a population of at least 15,000–30,000 with a density of about 50 persons per square mile. It is strikingly evident that they did not suffer the same debilitating effects as others and this signifies secondarily the immunity evident in archaeological findings of RI 1000.

There is only one way in nature to accomplish this, and that is to acquire immunities through exposure/experience. The Narragansetts therefore must have acquired it at some earlier period and place in their history. Their antecedents must have resided where the diseases were present, i.e. Europe.

While tuberculosis could still kill them, it took longer because their bodies fought it off with the antibodies that had developed from some European incursion hundreds of years earlier. There is evidence of this in the skeletons of Indians who died in the 1600s.

Other diseases such as smallpox were more lethal. So were massacres more lethal. After the Great Swamp Massacre, December 19, 1675, so many Amerinds were killed and wounded just west of Pettaquamscutt that the Narragansett society collapsed. By morning after the attack, they were essentially decimated and psychologically traumatized. There are some 2600 descendants, however, who still reside in Rhode Island to this time.

There are few, if any, of pure blood remaining in the tiny reservation in the town of Charlestown, Rhode Island. The Narragansetts are now pretty well integrated both socially and physically and are mostly indistinguishable from Caucasians today.

Unfortunately, the few remaining Narragansetts rarely speak up for themselves very forcefully. Amerinds of North America generally had little interest in history. This is not surprising since their practice was to "forget" the deceased and never mention them again. Europeans are just the opposite and are noted for their sense of history and attention to genealogy.

The end of the power of Narragansetts came about suddenly through the Great Swamp War, also called the Great Swamp Massacre. Reviewing the war between the Indians and colonists, Canonicus, chief of the Narragansetts, had given his allegiance to the king of England and was at peace with the colonists. The Rhode Island colony received its charter from the king, and took only a late and reluctant part in the King Philip's (Metacom's) War between Indians and colonists in surrounding states.

In retaliation for the killing of a few colonists, the united Colonies formed an army to attack the Narragansetts. This army formed and marched through Providence and Warwick on their way to the Great Swamp. Not until their territory was actually invaded did the Narragansetts offer resistance. After a short encounter during a blinding snowstorm, the whites entered a stockade in which there were over 500 wigwams, and the real fighting began. The wigwams were set on fire and soon the whole village went up in flames. Squaws with babies led children to escape but were shot or knocked out. Many were burned to death and many were shot or severed by swords.

The Great Swamp Massacre killed a thousand Indians and about one third were burned to death. The whites lost from two to four hundred. This was like a signal to all the tribes, and a pitiless war was now started, a war that the Narragansetts felt was for their very existence.

All the settlements suffered, even Providence, the home of Roger Williams, the best friend the Indians ever had. Roger Williams and his family were unharmed, although they refused to flee to safety.

Four months after the massacre, King Philip (Metacom) was slain, and with these two powerful chiefs (Canonicus and Metacom) gone, the Indians became demoralized, and the colonists gained the upper hand in 1676. The Indians could no longer plant crops and their food supply collapsed.

Those Indians captured were slain or sold into slavery. After King Philip's War came King William's War in 1689, Queen Anne's War in 1702, King George's War in 1744, the Canadian War from 1755 to 1763, and Pontiac's war in 1763. These wars signaled the end of not only the Narragansetts but many other tribes as well.

Let us now proceed to look at perhaps the most compelling difference of all between Narragansetts and other Amerinds—genetics.

Part VI

Narragansett Genetics Show European Ancestry

A detailed discussion of the genetic factor; how and why this shows that Narragansetts and Vikings were at one time intimate.

Narragansetts Have Pre-Columbian Genetic Differences

Settlers at Plymouth wrote about an epidemic (either one or a combination of diseases) which had killed Indians shortly before they arrived. From November 1620 to November 1621, William Bradford and Edward Winslow recorded all the activities of the little colony that first landed at Cape Cod and then settled in Plymouth. In 1622, George Morton published the account in London and called it *Mourt's Relation*.

The Englishmen who preceded the Plymouth settlers had brought the plague when they negotiated with Massasoit. Massasoit and the Narragansett Indians were unaffected by this devastating epidemic, undoubtedly due to resistances developed through their earlier interbreeding with the Vikings. The devastation surrounding them is obvious in these excerpts from *Mourt's Relation*.

> *And whereas at our first arrival at Paomet (called by us Cape Cod) we found there corn buried in the ground, and finding no inhabitants but some graves of dead new buried, took the corn ...*
>
> *The ground is very good on both sides, it being for the most part cleared: thousands of men have lived there, which died in a great plague not long since[1615]: and pity it was and is to see so many goodly fields, and so well seated, without men to dress and manure the same. Upon this river dwelleth Massasoit: it cometh into the sea at Narragansett Bay.*

A good reference related to this is in *The New Land* by Phillip Viereck. He noted that the sole survivor of the 1615 epidemic was Squanto who was in England when this happened. He returned to Massachusetts to find his entire village and family dead.

Unfortunately, resistance to smallpox is not something that can be seen in the skeletons left for archaeologists to find. However, resistance to tuberculosis can be found.

For the first time in the long history of Vinland studies, an irrefutable factor has been established—genetic evidence of a pre-Columbian contact upon the eastern shores of North America. Some 17 skeletons of 56 excavated near Wickford, Rhode Island, were discovered to possess lesions on rib, hip, and spine bones resulting from the disease of tuberculosis. So far as is now known, this is the most intense concentration of tubercular cases upon the North or South American continents. This is because most people died quickly of tuberculosis, too quick to form reactions to the disease. However, these 17 people lived long enough for their bodies to mount a supreme resistance that showed itself by lesions that took months or even years to develop.

Most remarkable of all is that a three-year-old child did not succumb quickly. The child had enough inherited resistance to continue living for many months while lesions developed on bones. Unexposed children who are naïve to the bacterial infection usually die very quickly, often within days.

This shows that individuals of this northeastern woodland Algonquin group possessed a response at distinct variance with other American aborigines to the toxin *mycobacterium tuberculosis* when exposed to it by European settlers in the 1600s.

This response was typical of a European Caucasian population. It was absent in other American Indian populations of the early colonial era. This trait of partial immunity is established from three sources. There are direct remarks of early colonial neighbors, reliable census figures recorded at that time, and discoveries of evidence within a primary gravesite.

The Narragansetts possessed genetic and anthropological attributes linking them to Europeans. Some of these attributes may have been the results of exploration and colonization enterprises of Greenlanders/Icelanders a thousand years ago.

Narragansetts Resisted Tuberculosis Because of Previous Exposure

Records and comments of the earliest explorers in the Pettaquamscutt-Narragansett Bay area identified varied social and genetic differences within the Narragansett Amerinds. These include disease immunity long before much was known about genetics.

Additionally, scientific excavations of a cemetery of these Amerinds found a particular genetic trait that is almost certainly of European origin of many centuries standing. The most plausible explanation of this occurrence is that these people must have had some contact with European culture at some time in their history. Since I believe the area had been visited at the time of the Vinland voyages, it is my position that the Narragansetts descended from the union of some perhaps renegade males of the Vinland voyagers and local Native American Indians.

Archaeologists discovered unaccountable anomalies in these cemetery exhumations such as atypical manifestations on skeletons, occurrences of which can be understood as immunology adaptation or as an acquired genetic trait.

The adaptation or trait, appearing in cemeteries of people buried about 1660 A.D, can be developed genetically only in reaction to a mycobacterium pathogen over many centuries. This finding has been a subject of interest among medical investigators.

When any population is afflicted with a toxin, the population will, for survival, adapt physiological defenses to the toxin. When these defenses are successfully established within individuals, they can then transmit these physiological defenses to their progeny. Once this occurs, those defenses become factors of heredity.

Such a toxin to humankind is tuberculosis, which is believed to be a transmutation of a bovine form of tuberculosis. European populations began evolving their defenses at some ancient period of cattle domestication while the New World population, lacking cattle, did not do so. Consequently, when the two populations came into contact after Europeans arrived, American natives demonstrated a marked susceptibility to tuberculosis, as well as extreme sensitivity to other European diseases.

These diseases devastated populations with no previous exposure quickly or epidemically. Early medical and colonial literature remarked upon a difference in reactions between Indian groups.

I accept medical opinion that tubercular lesions on bones are a typical European trait but rare and nearly non-existent among Native Americans.

The Vikings appear to be the only group near Narragansett environs long enough to have interbred with them and left this genetic trait. It was passed on to their descendants for 500 years when they were again exposed to tuberculosis by immigrating Europeans after 1492.

Sir Arthur Conan Doyle had his intelligent fictional detective Sherlock Holmes remark to Dr. Watson in the 1892 thriller, *A Scandal in Bohemia*, "Once you eliminate the impossible, whatever remains, no matter how improbable, must be the truth."

Jonathan Adams teaches archaeology at the University of Vermont. In his class synopsis of 1999, he wrote about the archaeological history of Rhode Island:

> *This period begins with Verrazzano's written observations of his exploration of Narragansett Bay in 1524 and ends with Roger Williams' settlement at Moshassuck (Providence) in 1636 ... [In] 1616, when a severe epidemic decimated native populations living all along the coast from Maine to Cape Cod. The Narragansett were physically unaffected by the epidemic ... Because they escaped the epidemic that was so devastating to other tribes along the coast, their status (and subsequent power and influence) among other native tribes was enhanced.*

Old World inhabitants typically can withstand the normally fatal tubercular lung affliction long enough to inhibit the disease as it attacks several parts of the body. Attachment of TB lesions upon the ribs, spine, or pelvis is one such diversion, a syndrome very rare for anybody in the Western Hemisphere.

It was a newsworthy event when these tubercular lesions were found on skeletons in an archaeological excavation of a 17th century Narragansett cemetery. The site was called RI 1000 and it was located in North Kingston, Rhode Island, and reported by an archaeological team of Paul Robinson, Marc Kelley, and Patricia Rubertone in 1985. An overview of the burial is contained in "Preliminary Biocultural Interpretations from a Seventeenth-Century Narragansett Indian Cemetery in Rhode Island" by the three authors in William Fitzhugh, editor, *Cultures in Contact: The European Impact on Native Cultural Institutions in Eastern North America, A.D. 1000–1800*, Smithsonian Press (1985).

More details on the site are in William A. Turnbaugh's 1984 *The Material Culture of RI 1000, a Mid-17th Century Burial Site in North Kingstown*. Another account of the find is in Marc A. Kelley's 1991 "Ethnohistorical Accounts as a Method of Assessing Health, Disease, and Population Decline Among Native Americans," in *Human Paleopathology: Current Syntheses and Future Options*.

The presence of lesions on bones indicated that the bodies had mounted a resistance to the infection, which enhances survival. These findings suggest extended histories of exposure and endurance resulting from genetic evolution.

The Narragansett Amerinds among whom the "resistance" trait was detected numbered at least 15,000 persons judging by Verrazzano's accounts of the population in 1524. They could not have developed this reaction since Columbus' arrival a thousand miles away in 1492 and subsequent expeditions. The time span is too short to produce the genetic effects seen in those Indians buried around 1660. Therefore, the trait must have been produced many years prior to 1492.

The presumption follows that people with tuberculosis "exposure" reached this area long ago, interbred with natives, and passed on their genetic exposure and immunity. Further, it is more than likely that the relative immunity is quite complex, not relying on a single gene but upon interdependent combinations.

The progenitors must assuredly have, at some phase of their history, resided where the toxin, or a population already exposed, had existed. Thus, ancestors of these people must have previously resided in, or visited, the Old World and then migrated to the New World, carrying these defenses that evolved from a more remote epoch of their history.

The invaders assimilated into a large and dynamic Narragansett population, which expanded to some 15,000/30,000 according to a census about 1670 A.D.

How did this vestige cross an ocean to appear in this extraordinary population within an area only 15 miles square?

Contagions spread with such speed as to seriously reduce or even annihilate a multitude shortly after a first contact with explorers. Even at settlements a few years later, a pattern of depopulation was noted and tragically repeated all the way into the Pacific Archipelagoes throughout the "age of discovery." Our discussion concerns not the contagion but biological reactions to it.

According to immunologists and scientists, the defenses to TB require centuries to become fixed within a population. Once fixed, they are passed on as quickly as a host can reproduce. More compelling, this genetic trait is supplemented by literary and analytical comments about it.

There appears no proof, evidence, or any substantiated suggestions of other long-lasting invasions of Europeans early enough to account for these genes other than the Vikings around 1000 A.D. or shortly thereafter. I believe that archaeological findings demonstrate both the event and dating of incursion of a people similar in culture to that of Iceland and Greenland.

These facts impel me toward this deduction: The aboriginal Narragansetts and the populations of Iceland and Greenland are related. They share coincidental ancestry over the 18 to 24 generations of commingling in 660 years. Thus, the New World Narragansett population bearing this demonstrable genetic trait (TB resistance) is descended in part from those Iceland and Greenland explorers and emigrants whom we refer to as the Vinland voyagers.

I believe that there is sufficient proof that this southern New England coastline is the lost land of Vinland and that Leif Erickson's and Thorfinn Karlseffni's landing sites were at the coastal estuary of Pettaquamscutt immediately adjoining Narragansett Bay. I hold that Vinland is at long last recovered and I believe its site is at Pettaquamscutt, an obscure and hilly parcel of terrain in South Kingstown and Narragansett in Rhode Island.

TUBERCULOSIS IN AMERICA

George A. Clark *et al.* wrote a 1987 article of historical data on tuberculosis, its biological course and evolutionary history in "The Evolution of Mycobacterial Disease in Human Populations." They postulated that the Amerind susceptibility to tuberculosis depended upon exposure to these bacteria, which explains why tuberculosis affects some native populations more severely than others. They began with the widespread belief that native peoples are predisposed to tuberculosis because they were never in contact with this in their past. The paper documented lesions on bones resulting from previous exposure in earlier generations to this dreaded disease. The authors offered no explanations and seven peer reviewers strained unsuccessfully to explain the exception found in Narragansett skeletons.

However, a Dallas anthropologist, Wilson W. Crook with whom I communicated, submitted this brief response to the Clark's findings in a 1987 issue of *Current Anthropology*, p. 543.

> *As evidence with regard to the presence or absence of tuberculosis in American Indians "before contact" with Europeans, Clark's data from a Narragansett Indian cemetery in Rhode Island is of dubious value. Whether one agrees or not, one very positive school of thought maintains that Leif Erickson's Vinland, occupied intermittently by different expeditions over three or four full winters, was precisely in Narragansett Bay. The sagas record definite and repeated contact, both trade and conflict, with the native "skraelings." If this is possibly true, then any*

> *evidence of tubercular affliction in the Indian remains of the period A.D. 1650 might well have been the inherited result of European contact in that particular area long before A.D. 1020–30. The authors do not mention this possible "contaminating" influence.*

Jordi Gomez i Prat and associates wrote in 2003 "Prehistoric Tuberculosis in America: Adding Comments to a Literature Review." This paper stated that tuberculosis was a prehistoric presence in the Americas and concluded with the following statements.

> *The direct and frequent contact with the animals and the use of contaminated milk and beef were probably responsible for the introduction of tuberculosis among the human groups ... Most of the cases of tuberculosis are dated from the second millennium, after the Vikings with their domesticated cattle settled in colonies on the Eastern Coast. According to written documents, they exchanged goods, including cow milk, with Indian groups, and this fact forces us to consider the possible penetration of European strains of mycobacteria in North America previous to the 15th century.*

Perhaps it should be mentioned here that tuberculosis infection occurs primarily by inhalation of infectious droplets usually from coughing, which may contaminate the air in closed spaces for long periods. In areas where bovine tuberculosis has not been eliminated, the disease may occur by ingestion of raw or contaminated milk and meat or even undercooked meat of a tubercular animal.

The Prat paper identifies some South American incidences of tubercular skeletal lesions clustered in the deserts and highlands of Peru, Chile, and a single case midway along the coastline of Venezuela. There are few incidences of skeletal lesions north of the Mexico-United States border. Gomez i Prat *et al.* suggest that the domestication of llama was a possible vector south of the United States.

I mentioned earlier that living patterns changed in the Narragansett Bay area about 1000 years ago. The most likely time for these genetic traits to have been passed to Amerinds was during this period of abrupt changes. Deborah Cox, *et al.* in *The Public Archaeology Laboratory* discussed this in the 1983 article "An Archaeological Assessment of the Pettaquamscutt River Basin." The essential cultural "shift" occurring about a thousand years ago indicated a radical change in lifestyle. The natives had relied on growing corn or maize inland for some time and changed abruptly to a coastal dwelling and fishing lifestyle over 900 years ago, reported archaeologists. Some passages from that article are excerpted here:

> *The survey on which this report is based was undertaken because the Narrow River presented unique data on environmental and cultural processes that could not be matched elsewhere in southern New England ... Instead of a simple evolutionary scheme, the survey has produced spatial and temporal patterns that suggest a complex land use system, which operated along the coastal river only for specific episodes ... the area was effectively unoccupied prior to 4,000 B.P. (Before Present) and probably ... 1,000 years ago. At this latter time, the river was reoccupied.*

The researchers specify that this culture had been an "advanced" one. This author concludes that this "advancement was to varied land use patterns. This suggestion of a different lifestyle from other surrounding Amerinds supports my hypothesis that Narragansetts were changed by contact with the Vikings. Then further support comes from comments by colonial observers such as Roger Williams who remarked, "The plague did not seem to be among them." Additionally, the respect they had among other nearby Indian tribes was because they were healthier, "tougher," and did not succumb to diseases.

Even if Vinland voyagers brought the actual disease with them, tuberculosis is a disease of crowding and dark interior living styles where people are breathing, coughing, spitting, sneezing, and singing close to each other. The outdoor lifestyle of Amerinds might have restrained the disease from epidemic proportions until some other factor came into play with the later European invasions. However, the chances are that the Vinland voyagers were a pretty healthy group.

The arrival of later European invaders brought a new social condition that broke these barriers. Small enclosed missions and trading posts in Spanish/French areas and nascent towns in the English colonies created more of the conditions for disease to spread. The arrival of the invaders was the "trigger" for diseases including tuberculosis.

In most books and articles about tuberculosis, the discussions resolve around the bacterium itself. Little, if any, consideration is given to the interrelationship with the victim animals or humans. Can a microbe affect victims in different ways or does it affect everyone in exactly the same way? The more complex the form of life (man), the more varied the reactions to toxins. Some people have more effective physical, environmental, and emotional responses to disease. They react differently than those with fewer reserves or different conditions. These reactions would vary as they interplay with the bacterium over time. I believe this explains why we see varied susceptibilities in various populations in response to diseases.

Varied susceptibilities by populations to disease suggest different victim response rather than different strategies or different causative organisms. The matter of population variation in resistance to disease is not new. In the past, it was a subject of intense interest, being probably the first question asked of arriving ships, caravans, and freight trains from foreign climes. Recent interest resulting from powerful microscopes, DNA, and worldwide travel analysis tends to obscure the subject, yet it remains as real as ever.

The tubercular lesions at the RI 1000 site still stand, by the extensive pathology on bones as well as age of victims, as the most concentrated incidence of all those discovered anywhere in the New World, north, south, east or west. Do those lesions necessarily indicate a resistance to tuberculosis due to prior exposure with people who had an acquired immunity?

An article, "Linkage of Tuberculosis to Chromosome 2q35 Loci, including *NRAMP1*, in a Large Aboriginal Canadian Family" remarked on the genetic influence on resistance to TB. Authors Celia Greenwood *et al*, elaborated on how this occurred. They identified the genetic codes of at least one type of defense that occurs on the genetic strand of a chromosome.

Dr. Richard Bellamy wrote about "Genetic Susceptibility to Tuberculosis in Human Populations" in 1998 for the journal *Thorax*. He discussed why people react differently to TB and some important comments are excerpted here:

> *Why do some people succumb to the disease (tuberculosis) when most of the population can successfully fight off the tubercle bacillus? Over 100 years ago, it was recognized that there are racial differences in susceptibility to tuberculosis ... A population's resistance to tuberculosis is determined by its previous history of exposure ... Even among these [racial differences], host genes influence individual susceptibility to tuberculosis, but which genes remains elusive.*

There are so many things to consider about who can or cannot resist diseases but genetic traits certainly play a large part. Drs. Laurent Abel and Jean-Laurent Casanova wrote about that in "Genetic Predispositions to Clinical Tuberculosis" published in the *American Journal of Human Genetics* in 2000. They said:

> *Several studies have shown that a person's resistance level to Mycobacterium tuberculosis correlates with the region of his or her ancestry and that the ancestors of more susceptible persons tend to come from areas once free of tuberculosis. Similarly, the incidence of tuberculosis has been found to be particularly high during outbreaks in populations, such as Native Americans, with no ancestral experience of the infection.*

The peculiar presence of tubercular lesions on bones in Narragansetts indicated an admixture whose evolution by 1660 A.D. must have consisted, in part, of European genes.

Another article I found valuable was a 1992 article called "Telltale Bones" in *Archaeology Magazine* by Clark Spencer Larsen. The article evaluated over 500 exhumations of Guale Indians of Amalie Islands of Florida. They exhumed human bones below a church and other structures destroyed by fire in 1597. Therefore, the bones were older than 1597. Tuberculosis exposure as indicated by skeletal nodules was specifically sought and found absent in that population. Thus, despite a similar lifestyle and colonial contact, they did not have the evidence of tuberculosis found in the site of RI 1000 of Amerinds who died before 1660.

Still another interesting book is *Yankee Surgeon: The Life and Times of Usher Parsons, 1788–1868,* by Seebert J. Goldowsky, M.D. Dr. Usher Parsons had been the sole surgeon of the American Fleet at the Battle of Lake Erie in 1813. Parsons, by the way, was the brother-in-law of Oliver Wendell Holmes, M.D. He later settled in Rhode Island and took exceptional interest in the Narragansett Tribe. He excavated numerous Indian graves and one contained a known Narragansett "princess" or "queen." Parsons reported on the excavation in an article and at historical societies.

Therefore, we cannot be certain that this Narragansett woman was a victim of tuberculosis. However, we shall return to Dr. Goldowsky, Parsons' biographer, as he described what Parsons and others found in the Indian graves, p. 406–407.

> *In May of 1859 several citizens of Charlestown, Rhode Island, spurred on by curiosity as to whether Indians were buried in a horizontal or sitting posture, repaired to the Sachem's burial ground with spade and crowbars ... The body [buried 12 feet deep in a split log sarcophagus]when revealed, was elegantly arrayed in a green silk robe. A silk cloth covered the head, and from it descended a silver chain to the soles of a pair of copper moccasins lined with leather. These were of neat workmanship and indicated a slender and delicately formed foot. Around the waist was a belt covered with bead work of wampumpeague (Indian shell money) and studded with silver brooches ...*
>
> *Two loose coins were found, one an English farthing and the other a French silver half livre, scarcely worn at all, dated 1650 and bearing the legend Ludovicus XIIII (Louis XIV), whose reign began in 1643 ... and she was buried in the first grave in the Sachem's cemetery, indicating that hers was the first death in the family ...*
>
> *Usher was able "with some pains and trouble" to recover a skull and skeleton from the next adjacent grave. The bones conformed to an age of seventy, as did the*

angle of the lower jaw, as confirmed "by an eminent dentist." The os femoris (femur, i.e. the thigh bone) indicated a man of large stature, more than six feet tall. This was undoubtedly the skeleton of Ninigret himself, Sachem of the Niantics, and the smaller one with its elaborate array of relics that of his young daughter.

An eminent and industrious Rhode Island historian, Sidney Smith Rider, attended a lecture by Dr. Parsons conducted during America's Civil War (October 7, 1862) which had, as an interesting part of the discussion, a skull of a Narragansett female known to have been a daughter of the famed Chieftain Ninigret. Both these remains and also Ninigret himself had been excavated from the burials at Charlestown in May of 1857, some five years earlier. Rider took interest in this skull and examined it closely, turning it about in his hands after the lecture. He noted a clot of what he thought was dirt and for some reason scraped this dirt off and preserved it upon a microscope slide.

For quite some time, nothing occurred. But in 1892, he read a paper that stimulated his curiosity in this matter and he took up the slide, re-examined it, and wrote thusly of his observation through the microscope. (These observations were made by a man who had been active in Rhode Island history in all aspects and who at the time was well into his seventies.)

On the evening of the 7th of October, 1862, nearly 50 years ago, the Rhode Island Historical Society held a meeting to listen to a paper from Dr. Usher Parsons about the opening of a grave. I was a member, and present; there were about 18 members present. I was the 18th—not one of these members, myself excepted, is now living. I took up the skull and found on the back of it a small bunch of what I took to be earth, but it proved to be filled with short pieces of black hair; with my knife I lifted a bit and carried it home. I brought my microscope to bear upon it.

It was a bit of hair. I mounted it in Canada balsam on a slide for use under my microscope, which I was then much interested in using. I can show that the hair was jet black like that of all Indian women. I can show that it was tubercular. I have the slide and the microscope to this day. All this came to me when I read ... Tucker's History, and I rushed at once to find the skull. But nobody had ever heard of it. I ran, actually ran, to Rhode Island Hall [at Brown University] to see their skulls. Dr. [Albert Davis] Mead [of the Brown biology department] led me to them, and there I found one beneath which was a card on which was written. "Skull of Sachem Ninigret's Daughter, Charlestown. Prof. C. W. Parsons, M.D. see Historical Magazine, Feb. 1863." It was the skull of Wuenquesh, the daughter of Ninegret; she became queen of the Niantics upon her Father's death in 1676."

In my opinion, this observation is exceptionally important despite the loss of provenance and some few understandable errors. It requires some examination. The main thrust of interest is the excitement of Mr. Rider who, at age of over 70 felt such inner turbulence that he ran to see without delay the skull he had observed 50 years before but which more recent information had made new to him. It must be noted that interest in communicable disease was much more prevalent in those days than it is now when people depend upon public health authorities. Nearly everyone was alert to disease symptoms in those times of high infant mortality and when quarantine had to be instituted early and earnestly by family heads and mothers. And, in fact, a number of historians recorded that even Native Americans individually were alert to varied disease symptoms which might indicate communicable or otherwise.

Tucker's History, mentioned in the quotation above, is from William Franklin Tucker, *Historical Sketch of the Town of Charlestown in Rhode Island, from 1836–1876* written in 1877.

The skull Rider examined at the biology museum at Brown may not have been the same as that he examined in 1862, for the description of the quite opulent 1857 exhumation says that the body had been interred some 20 years prior to Ninigret. But the skull in the quote was identified as from another daughter who became Queen in 1676. Whichever person it was, she was certainly a relative by blood of the Narragansett (Niantic) Chieftain Ninigret. The Niantics were a sub-tribe located west of Narragansett only some 10 miles away.

The skull had been donated to Brown University by C.W. Parsons who was Dr. Usher Parson's son. The misidentification of the skull, if it occurred, must have been a result of passage of time and confusion at inheritance.

However, with all this I must confess that modern medical literature commentary on tubercular hair seems so sparse as to make the symptom seem of little consequence. However, it does appear with some frequency in papers and publications concerning hair pathology where I have seen it described as hair with a thick yellow crust and with white spots along the strands. It would be strange, indeed, if TB did not have an influence upon growing hair, for it was recorded that in addition to its internal pathologies it also has observable effects upon both skin and fingernails.

Parsons, a native of Maine, became a surgeon in the new U.S. Navy and his first assignment was to the ship *U.S.S. Lawrence* which was the command ship in the battle of Lake Erie on September 10, 1813. *U.S.S. Lawrence* was the flagship of the Rhode Islander Commodore Oliver Hazard Perry, well known for his famous saying: "We have met the enemy, and they are ours."

Dr. Parson's "hospital" was a room only 9-feet square and happened to be precisely upon the water line level, which had become the main aiming point of the combatant British fleet. No fewer than eight cannon balls crashed through the tiny room while Parsons was attempting succor to many wounded. Two hit his patients while being treated. One of these two was Charles Pohig, a Rhode Island crewmember who was a Narragansett Indian. Parsons and Pohig were shipmates and must have known each other during the months before the battle. It may have been from this encounter that Usher Parsons also took exceptional interest in the Narragansett Tribe after he established practice and residence in Rhode Island.

The protracted history of tuberculosis in dynamic Europe, that hotbed of war, trade, and population movement, systematically reduced susceptible individuals and left hardier folk to reproduce. This had the effect of eventually producing whole races of individuals with the ability to sustain attacks of TB, and surviving a lung affliction long enough to sustain secondary symptoms such as indicators on bones.

The RI 1000 excavation of the 59 graves buried from 1650 to 1670 was described in *New York Times,* September 26, 1983, in an article entitled "Indian Remains Shed New Light on Syphilis." The newspaper article was not as complete as later archaeological reports but an excerpt from the *Times* describes the event.

> *The discovery is "very significant," Professor [Marc] Kelley said … The bones of at least five Indians also showed evidence of tuberculosis and there was some indication of smallpox … Edward F. Sanderson, deputy director of the Rhode Island Historical Preservation Commission, said discoveries there had shaken notions about how the lives of the Narragansett Indians were influenced by European settlers.*

Professor Kelley said 60 percent of the Indians buried there were 3 to 17 years old.

The excavation showed that "nearly all" of the burials had bodies placed in fetal positions on their right side. The graves were compared with certain graves of Vikings in Iceland and Sweden whose bodies were placed in that manner. The cemetery layout was neat, geometric, and orderly. It was unique in that it was long and narrow with bodies four abreast. It was not Christian, neither was it in imitation of Christian cemeteries. All bodies in RI 1000, as the cemetery itself, were oriented toward the southwest. (The southwest quadrant had ritual significance within the cultures of both Narragansetts and Vikings as I mentioned earlier.) Grave goods show unexpectedly intimate contact with English and Dutch colonials.

William H. McNeill's *Plagues and Peoples* delves into spans of historical records examining public health records from the past. He was able to determine factors of plagues and pestilences that have afflicted populations over eons. Consequently, he has discerned particular diseases and their courses in eras and severity. His statistical base enables us to understand diseases and immunities in a way not usually approached by professional scientists or immunologists.

He found that certain disease epidemics recur in cycles and that within these cycles specific diseases tend to follow paths of diminishing effects. For example, when a disease first strikes a population, it does so with terribly grotesque and "florid" symptoms, affecting individuals traumatically, with immediate depletions in numbers. However, as a disease cycles through long-term epidemic/endemic patterns, its effects diminish. The population and the attacking microbes slowly reach an accommodation with each other, responding in typical patterns as each organism gains experience and sometimes strength.

At every epidemic recurrence of a disease, symptoms become systematically less and less severe. What starts at first strike as severe disease with extremely traumatic effects, progresses to one less traumatic, to become a relatively minor affliction or a childhood disease and may ultimately disappear. The two formerly hostile organisms reach an eventual accommodation to each other with eventually no harmful effects observable. Diseases like measles and mumps are now, in Caucasians, mild childhood diseases but were formerly quite severe. Their severity can be seen in the mortality when contracted in adulthood by those unlucky in the inheritance draw or in certain non-Caucasian populations, particularly Native Americans.

Tuberculosis has always been a subject of alarm and consequent attention by public health officials. Nevertheless, McNeill remarks that tuberculosis is a very old disease. Tuberculosis is a transmutation form that has crossed from cattle to humans. Consequently, the immunology cycle of diminishing effects commenced at about the time of cattle domestication. This must be an "Old World" phenomenon, since "New World" aborigines did not domesticate cattle. They did, however, domesticate the ruminants llama and alpaca, and the rodent guinea pig.

Therefore, we can establish a time differential. That is because the indigenous populations of the New World descend from a population that crossed the Bering Strait at some time prior to cattle domestication.

Thus, epidemic/endemic cycles of TB have progressed for a considerable time, perhaps 10,000 years, and there are "Old World" populations who have developed partial immunities to this disease. In its simplest form, it is a contagion of the lungs, which secondarily results in debilitation of the body, with death resulting in a relatively brief time. It is also known as "consumption" because of this wasting away of physical resources. Longevity depends upon the genetic history of the victim and whether there has been a history of exposure.

So diverse are morphologies of "Old World" responses to tuberculosis that some say it is a difficult disease to contract. Others aver that it is so contagious that it can be contracted from droplets in the air, which happens to be true. In fact, both observations are accurate, depending upon which population is being considered. This difference is explained by the historical experience of the population of the victims.

It is claimed by some specialists that before antibiotics, 95% of Caucasian children had been exposed to TB by age ten but that the majority succeeded in resisting the disease with little further risk of contagion.

It was found that a small percentage of children developed "spots" (scars detected by x-rays) on the lungs from a less than successful resistance and that these individuals were at great risk of contraction or re-infection at a later period of life. In fact, I have known several friends where this was the case and they died later as a result of re-infection. Often such people are suspected to be "carriers"— people who did not demonstrate overt symptoms but who transmitted the disease to others. Public health authorities made every attempt to isolate these people by removal from general school populations, etc.

There are "Old World" populations where the epidemic/endemic cycles are so extended that many individuals can survive TB and live to old age, diverting secondary symptoms to other parts of the body, and maintaining energy enough to reproduce themselves and thereby contribute their biological strength to their progeny. European history and literature is replete with the histories of persons who lived relatively normal lives with what was termed "consumption" and certain other disorders.

Moreover, in the order of things, if one parent is weak in resistance, it may happen that the strengths of the stronger parent may prevail in the genetic order of three to one. Offspring failing to inherit the strengths will therefore, being susceptible, be more likely to contract the disease and either die or lack resources to reproduce. In this way, a population will tend to gain resistances to disease precisely as it does to any other inimical attack.

Tuberculosis is characterized by the formation of small "tubercles" which initially affect the lung, but which, if death is deferred, then migrate to other parts of the body such as neck glands. If the victim survives long enough, it may happen that these tubercles will attach themselves upon certain bones, usually the ribs near the lungs and the spine, and then lower in the pelvis. When this happens, they often deform the bone and leave characteristic lesions, the presence of which is one post mortem symptom of tuberculosis.

Immunologists term this sort of circumstance as effects upon "naïve" populations—those who have not had contact with a disease. "Sophisticated" populations are those who have had at least some contact with the disease and some evidence of it even if only in their genes. Europeans, in general, were "sophisticated" for TB since just about all of them practiced cattle domestication for a long period of their history and had therefore been exposed.

Diseases introduced by Europeans were responsible for vastly more death and destruction of Amerind societies that lived in close quarters. The terrible and tragic human depletion is only now becoming understood. Some estimate that at least 95 percent of Native Americans were eradicated as these diseases extended across the South Seas all through the "Age of Exploration." As comparison, the Bubonic Plague, thought to be a horror of the ages, only took something like 30 percent of the Old World population.

Indians everywhere were victims of "first strike" epidemics, which often reduced them to small percentages of former numbers. Ordinarily, if left alone, they might have recovered their numbers with newly introduced immunities. However, in many cases, while so reduced, European settlers appeared bearing infected blankets, other poisonous items, and antagonistic attitudes. Those re-infections had the result that few tribes recovered or prospered and many were eliminated.

New attitudes and new medical approaches appeared in the early 20th century and now the Indian, while still more susceptible to disease than whites, has only a much less antagonistic social prejudice to contend with and his own internal differences to resolve.

How long is the epidemic/endemic cycle of tuberculosis? American Aborigines have developed little resistance, if any, to TB since the Europeans settlers arrived. Probably we might presume a timeline in the order of 10,000 years since cattle domestication began in the Old World.

Today, Native Americans who survive TB are beneficiaries of modern antibiotics and not any genetically acquired resistance. They suffered tuberculosis acutely from about 1492 to the present in a distinctly "naïve" manner; i.e., as a pure lung affliction resulting in death for adults in a year or two and children but a few weeks or months. Just after contact with the first explorers, the "first strike" of "florid" effects could kill some in only a few days.

The appearance of these tubercular lesions among Narragansetts in the New World shows that those individuals were descended from a line that had developed partial immunity over a long time. There simply is no other way for these lesions to appear except by contraction of the disease and membership in a gene pool that has developed some relative immunity.

The great populations remarked by Cortez in central Mexico were unable to resist TB. The presence of the occasionally reported case of TB in Peru, therefore, might be a result of contact with llamas and alpacas. TB does move through different populations and since cattle introduction to the New World, TB is now found in some deer, elk, fox, badgers, and even a bear.

We see from several documents that Narragansett public health was much better than their neighbors were, and seemingly parallel to that of incoming colonial populations who possessed long-term contact with tuberculosis. The census of 1670 shows their population not only sizeable but also compact and dwelling in such close quarters as to seem especially susceptible to epidemic diseases. Two knowledgeable observers remarked that the plague did not seem to be among them at that time.

The tribe was able to mount a sizeable army in 1676, a century or more after contact with various explorers and settlers. That was at a time when most other aborigines of the area were decimated or annihilated. An army of at least 1,500 perpetrated the attack upon Providence during 1676 and some researchers estimate nearer 2,000. This was not only the largest body of men formed as an army in New England by either Indians or Colonists, but it may have been the largest ever formed into an entity by any Indians north of Mexico ever. This was symptomatic of both health and hierarchy. Narragansetts, alone among Amerinds, could be "drafted" (forced, conscripted) by their "princes." That was at a time when all other Amerinds in contact with Europeans were dying and declining.

Narragansetts stand alone among all "New World" residents, inclusive of South Sea Islanders, in "escape" from some European pestilences. This is a signal of varied genetic makeup. Summarizing critical references, Clark establishes a factor, Prat expands and delineates the issue, and Cox defines an incursion in Rhode Island and its dating.

Many historians propose that the Vinland voyagers were driven away or became discouraged by ill relations with the aborigines in residence. Documents show most of the 200 or more Vinland voyagers were male. Some, at least, had little reason to return to their homelands of Greenland and Iceland. For some, it was more desirable to stay in America and "go Indian" or allow Indians to "go Viking." Native American females could tend hearths, cook, dress furs and skins, and rear progeny.

Their progeny were noted in the anomalies of land use distinctions in the Pettaquamscutt River Valley, which commenced about a thousand years ago. They possessed language variations, cultural differences, adaptations to natural conditions as their native mothers, and the genetic traits of their adventurous fathers.

Let us pursue the theme of intermarriage between Narragansett Indians and the Norsemen. Would that account for Giovanni da Verrazzano's comment that the Narragansett visitors sent their women to a small island rather than risk boarding of the ship *Dauphine*? They likely had legends of what had happened when earlier voyagers came amongst their people.

In these many long years of study, I believe the Amerinds early on understood that the white man brought diseases and sex-starved men that would ultimately destroy them. However, their efforts at resistance and cooperation came too late and likely would have been futile no matter what they did.

A fine Amerind insight into the effect of disease comes to us from *Madonna Swan—a Lakota Woman's Story* as told to Mark St. Pierre, published in 1991. The Lakota are a Plains tribe dwelling essentially in the area of North and South Dakota, Wyoming and Montana. Some segments of the tribe are called Sioux. They are famed as heroic warriors and fiercely independent people, largely responsible for overcoming General George Custer at the 1876 battle of the Little Big Horn in Montana.

While Mrs. Swan's story is of modern times, it does span a period to the past. Her uncle (or great uncle) was present at the Little Big Horn where he aided women and children to escape what was initially feared as an approaching massacre. She was born in 1928 on the Cheyenne reservation in South Dakota. It was her misfortune to contract tuberculosis at age 15. She was hospitalized for ten years and was saved by draconian treatments of removal of all her left ribs and left lung. She survived and eventually became a teacher and respected person among her people.

What she has to say concerning TB and her experiences is of interest to us. First is the important element that the Indians called tuberculosis "the white man's disease." Mrs. Swan describes several types affecting her friends of which many of us had been unaware. Primarily it infected the lungs as it did Mrs. Swan, but she described several other types such as TB of the skin (frequent eruptions resulting in death). That form of cutaneous tuberculosis was discussed when I dealt with the effects upon hair in the Indian princess or queen examined by S. S. Rider.

In her book, I learned that contagion was more rapid than would be expected in a Caucasian environment, often resulting in death within two years from onset and in children much quicker. She described several episodes where death occurred a mere two days after onset of symptoms.

Her ordeal was long and painful and she probably would have died if she had not been hospitalized. Five years after her admission, she was transferred to a regular (Anglo) sanitarium where more advanced treatments were available, as well as the extreme operation of removal of ribs and lung. During her stay, her younger brother, also a victim, joined her. I cannot determine how long his pathology extended, but from the way she describes it, he seems to have died within a year.

She and a friend both kept diaries. When they compared these diaries some years later they noted that from 1944–1945 no one was discharged from the hospital at all. Death was the only exit. In 1950 the two recorded 500 deaths and this was a count only of those they were aware of. It did not include those they had not heard of and did not include those who went home and later died, or those who ran away and died. This was a span of some five years in a hospital that served a total population of perhaps 50,000 people. This death rate is very high and could explain a severe reduction in overall population from TB. The deaths are one thing, but the effects on survivors are another. Mrs. Swan had nine siblings. Five died young, only one from an accident and the rest must have died from disease of some sort.

The Web site of the American Lung Association states that the rate of tuberculosis per 100,000 is nearly six times greater in American Indians/Alaskan natives than in whites. This shows the difference between those who have been exposed to the disease via a European ancestry and those who have not. Narragansetts were an exception to the multiple studies about how Amerinds succumbed to the disease.

The Amerind vulnerability to disease is described in "Virgin Soils Revisited" by David S. Jones published in 2003. Jones commented,

> American Indians suffered terrible mortality from smallpox, measles, tuberculosis, and many other diseases ... William McNeill and Alfred W. Crosby in the 1970s both argued that the depopulation of the Americas was the inevitable result of contact between disease-experienced Old World populations and the "virgin" populations of the Americas ...

And yet—Narragansetts escaped! This work attempts to explain why.

Would DNA Tests Determine Viking Descent?

Many think this whole issue of whether the Narragansett Indians have Viking blood might be resolved by DNA tests of current Narragansetts. That is not as simple as we might like to think. Consider this information by Deborah A. Bolnick of the University of California. She presented a paper called "Showing Who They Really Are" at the American Anthropological Association Annual Meeting on November 22, 2003.

She said there are three different types available—mitochondrial DNA tests to trace maternal lineages; Y-chromosome tests and a test of genome-wide markers. She described the problems such as using the modern-day Norwegian population to represent the Norse Vikings and said the tests ignore the variation found within all human populations. She added that the results for particular individuals have changed significantly over the past year. They are often being used to obtain race or ethnicity-specific scholarships, commercial opportunities, government entitlements, and college admissions.

In view of this information and similar opinions from other scholars, science and DNA have not advanced enough to conclude whether there is Viking blood in today's Rhode Island Amerinds.

An important article about how a small group might possibly increase over 500 years is "Inbreeding effects on fertility in humans: evidence for reproductive compensation" by Carole Ober *et al.* in 1999 in *The American Journal of Human Genetics*. Hutterites, whose origins were in the Tyrolean Alps in the 1500s, were attracted to the promise of religious freedom in Russia in 1770. There their population grew from 120 to 1000. In 1870, some 900 immigrated to the United States in South Dakota. About half settled on communal farms and the other half on single family holdings nearly 130 years ago. The Hutterite population is now 35,000, and all extant Hutterites can be traced genetically to less than 90 ancestors who lived during the early 1700s to early 1800s.

Their situation is not unlike the Vinland voyagers, some of whom must have stayed on in Vinland. The increase in 130 years speaks for itself.

Suppose we compared Nicaragua and Nebraska. Visitors to these places would immediately observe many genetic differences between the two peoples. However, something unobserved is the invisible existence of sophisticated resistance to the dread disease Yellow Fever. That illness, now under control, is "endemic" in Nicaragua and non-existent in Nebraska.

If we could test the entire population of Nicaragua, we would find specific "antibodies" in the blood of nearly everyone there because they have a long-lasting experience with that disease. However, suppose we could test the entire population of Nebraska. The antibody for Yellow Fever should be non-existent. If it were to be found, immunologists would immediately assume that at some time in the past, that individual or maternal ancestors (since antibodies are transmitted through colostrum or mother's milk) had visited the area where Yellow Fever was endemic.

The disease is area-specific and antibodies cannot be formed unless the individual has some contact with the causative microbe. The antibodies are formed within an individual's body as a defense against the disease. Inoculations are effective by introducing an antibody or by injecting a benign form of the disease to stimulate the manufacture of antibodies in the bloodstream. Once the antibodies are present, that individual is thereafter immune to the disease.

He or she might have contracted it and survived, or simply contracted it and was successful in resisting it. A female will transmit the resistant antibodies in her mother's milk (colostrum) during her first three days of nursing.

It is obvious that the people of Nebraska cannot develop immunities to Yellow Fever without either traveling into the endemic area or being inoculated against it. Nebraskans simply do not and cannot possess immunities to this disease since it is not present in their area. It simply cannot happen.

In dealing with tuberculosis, however, while the issue is a bit more complicated, the principle is the same. The introduction of tuberculosis to the Americas was around 1492 A.D. or earlier or later. New World natives, except the Narragansetts, did not have adequate personal defenses when they ultimately did become exposed.

Time is something we may regret when we look in the mirror or observe our childhood friends. Yet it has its values. Things change. We learn. We are not the people we were a few years ago. Indeed, we are not really the same as we were a few hours ago. For in that brief time the perpetual battle of microbes has been waging within our bodies. Our somatic defenses are in constant reorganization to new microbial and viral threats of which we are unaware. Perhaps all those white blood corpuscles are marching now to a different drumbeat; to a better-dressed line, with drillmasters that are more experienced.

I realize that the Narragansett Indians were extremely lucky in having a resistance to TB by some prior interbreeding with people who had been exposed to the disease in Europe. Other Amerinds were not as lucky and succumbed to the disease quickly as they did to so many other diseases brought by Europeans.

However, genetic studies have also contributed to finding the origin of diseases caused by the Vikings. Dr. C. M. Poser used such studies in his 1995 article called "Viking Voyages: The Origin of Multiple Sclerosis." He summarized his results:

> *Multiple sclerosis is most frequently found in Scandinavia, Iceland, the British Isles and the countries settled by their inhabitants and their descendants, i.e. the United States, Canada, Australia, and New Zealand. This suggests that the Vikings may have been instrumental in disseminating genetic susceptibility to the disease in those areas, as well as in other parts of the world.*

Genetics can help us learn more about our origins in the future. In recent years, genetics have advanced and we can determine where some people come from using DNA and X-Y chromosomes. For example, a study by Helgason *et al.* in 2000 found that 20–25% of Icelandic founding males had Gaelic ancestry and 75–80% had Norse ancestry. The researchers concluded in the *American Journal of Human Genetics* that

> *… numerous slaves were captured by the Vikings in their raids on the British Isles, and many of the slaves were taken to Iceland. The majority of these slaves seem likely to have been female.*

This confirms my theory that everywhere Vikings went, they left their genes, even among the Narragansetts.

In these 30 long years of study, I have found it extraordinarily difficult to learn the Amerind point of view concerning these issues. This is primarily because there was no universal or written communication throughout the Americas at the time of first contact.

Returning to my original quest, Leif Erickson's landfall has been sought by many over the centuries. Each explorer has been faithful to the information available to them. It was my advantage and good fortune that I was given more time and more resources than many of my predecessors.

The common belief that Vinland settlers were driven off by ill relations with native inhabitants may be no more than just a mistaken belief in view of the information heretofore presented. Perhaps not all Vikings were discouraged enough to return North. Perhaps in the years before some departed, Scandinavian seamen had found native women to share their lives. Perhaps some remained behind from choice, for this was the way of the Viking. Viking crews shed disgruntled or disaffected seamen wherever seamen went, leaving surprising legacies in genes everywhere.

Who Else Could Have Interbred with Narragansett Indians?

We know that the genetic resistance to tuberculosis means that the Narragansetts interbred with some Europeans long before the 1600s. Are there any other possibilities other than the Vikings? Let us review who was in or near Rhode Island prior to the first European settlers about 1639. Groups that might have interbred with Narragansett Indians could have left genetic traits in the 120 years or so after Columbus' voyages and the first plague in 1615 in the Rhode Island/Massachusetts area. Much of the following information about early explorers and their visits to this area comes from the work of esteemed historian and seaman, Samuel Eliot Morison.

Timeline for the Discovery of America before 1670

500–800 A.D.	Irish monks such as St. Brendan sailed a northern route to America in small ox-hide vessels called "curraghs." There is no evidence that they got to Rhode Island.
800–1347 A.D.	Vikings explored and set up established colonies in the Faroes, Orkney, Iceland, Greenland, Labrador, Newfoundland (L'Anse aux Meadows Viking settlement)

and Vinland. We suspect they got to Rhode Island and stayed in the Americas several years.

1398 A.D. Henry Sinclair, "Prince" of Orkney Islands, sent Nicolo Zeno to survey Greenland to prepare for a journey to the New World with 12 vessels of monks and fugitive Templars. Some think Sinclair then sailed south to Rhode Island. Proof is tenuous.

1492–1502 A.D. Christopher Columbus reported to his son that he traveled as far as Iceland before he made his four historic voyages discovering the Bahamas, Hispaniola, Cuba, Jamaica, Panama, and other islands. He was never close to Rhode Island.

1496 A.D. John Cabot, a Venetian, sailed from Bristol, England, to the northern tip of Newfoundland, to its southern coast and back for Britain's King Henry VII but he did not sail south of that point.

1500 A.D. Joao Fernandez from the Azores sailed to Greenland for the Portuguese in search of Cathay but did not sail south of Labrador from all accounts.

1500 A.D. Gaspar Corte Real sailed for Portugal to Newfoundland (according to Morison). He sailed again May 1501 but was lost. Two of his ships returned in October 1501. There is no definitive evidence that he was in Rhode Island.

1502 A.D. Gaspar Corte Real's brother, Miguel, sailed in May 1502 to find his brother. His ship was also lost but one of his fleet returned. There is no definitive evidence that he was in Rhode Island.

1504 A.D. French Norman fishing vessels were recorded on the Grand Banks of Newfoundland and continued for hundreds of years, joined by Bretons, Basques, and others. Some appear to have fished in Rhode Island but there is no evidence that they settled there.

1520–25 A.D. Joao Alvares Fagundes of Portugal sailed to Newfoundland and into the Gulf of St. Lawrence first. He later set

	up a colony to make soap and cure codfish on Nova Scotia's Cape Breton Island at Ingonish but was driven off by hostile Indians within 18 months. There is no evidence that he came as far south as Rhode Island.
1524 A. D.	Giovanni da Verrazzano explored the Atlantic coast for six months from Cape Fear in North Carolina to Newfoundland. He spent two weeks in Narragansett Bay but Indians kept their women away.
1524–1525 A.D.	Portuguese sailor Estevan Gomez sailed for Spain to Nova Scotia's Cape Breton, entered the Gulf of St. Lawrence, sighted Prince Edward Island, sailed for 11 months to Bangor, Maine, and Cape Cod. He kidnapped some Indians for slaves that could have been Narragansetts but was in that area only briefly.
1525 A.D.	Luis Vasquez de Ayllon sailed for Spain with three ships, 500 people, friars, black slaves, and 80–90 horses to found a colony along the Carolina coast around Wilmington. Within a year, 150 survivors returned to Spain due to internecine battles and hostile Indians. There is no evidence that he was in Rhode Island.
1527–1528 A.D.	John Rut sailed from England to "discover the land of the Great Kahn." He found a cape on Newfoundland, Cape Breton in Nova Scotia, fictional "Norumbega" in New England, the West Indies, Santo Domingo, and Puerto Rico before returning to England in 1528. While the "Norumbega" he described was not far from Rhode Island, Rut's voyage in the area was short.
1534–1536 A.D.	Jacques Cartier sailed for France to Newfoundland, landing first at Cape Bonavista and then only a few miles from L'Anse aux Meadows, the Viking settlement, at Quirpon. He sailed down the west coast of Newfoundland, Prince Edward Island, and Gaspe Bay for five months. He sailed for 14 months again in 1535 exploring the St. Lawrence River, founding Quebec and Montreal, but was not near Rhode Island.

1536 A.D.	Richard Hore sailed from England to catch codfish and to give London gentlemen a pleasure cruise. They landed at Penguin Island off the south coast of Newfoundland. Short on supplies, they resorted to cannibalism but captured a French ship and returned to England seven months after their departure. There is no evidence that they were near Rhode Island.
1541 A.D.	Jacques Cartier sailed with five ships from France to set up a colony near Quebec. He took many criminals because others were afraid of such a voyage. Manon Lescaut, an 18-year-old female prisoner was aboard, generating a book and opera about her. Hostile Indians and scurvy forced them to leave in 15 months and there is no evidence that they were near Rhode Island.
1576–8 A.D.	Martin Frobisher explored Frobisher Bay and believed (wrongly) that he found gold there. He made three short annual visits but never found the Northwest Passage to Cathay or valuable ores. There is no evidence that he was near Rhode Island.
1583 A.D.	Sir Humfrey Gilbert sailed with 260 people from England intending to establish a colony in Rhode Island and Connecticut. He ran into bad weather and his few survivors returned after four months. He did claim Newfoundland for England but never got to Rhode Island.
1585–7 A.D.	John Davis made three short trips from Britain to find the Northwest Passage to China and stopped at Greenland at Davis Strait. There is no evidence that he was near Rhode Island.
1607–10 A.D.	British captain Henry Hudson sailed to find the Northwest Passage but found Hudson's Strait and Hudson's Bay before his crew mutinied on his last trip of 1610–11. There is no evidence that he was in Rhode Island.

Who else was in America before the first settlers in Plymouth Colony and in particular Rhode Island in the early 1600s?

Ireland was the center for a vigorous culture from 500–600 A.D., preserving Christian civilization in Northern Europe after the decline and collapse of the Roman Empire. During this period, Irish monks sailed the North Atlantic in pursuit of some kind of spiritual or divine mission. They reached the Hebrides, Orkneys, and Faeroe Islands. Obviously, without some women to yield progeny, their settlements could only last as long as each individual lived. Maybe they lost their faith occasionally, just long enough to reproduce.

The Norse sagas suggest that Irish monks were in Iceland when the Norse settled there around 870. Those sagas add to the story of St. Brendan, born in Ireland about 489, who founded a monastery at Clonfert, Galway. He and 17 other monks set out on a westward voyage in a wood-framed boat covered in sewn ox-hides. The monks sailed about the North Atlantic for seven years, according to the *Navigatio Sancti Brendani Abbatis* written in the tenth century. Eventually, they reached "the Land of Promise of the Saints," which they explored before returning home with fruit and precious stones found there.

If Irish monks did voyage across the Atlantic and back, then their achievement was historically very significant. Ireland was the target of Viking raids before the end of the eighth century, and it is perhaps through the Irish that the Norsemen learned about other lands further to the west.

In 795 A.D., Culdees or Irish monks appeared in Iceland. They wore white robes and used bells, books, and croziers. They marched in procession, bearing banners on long poles and shouting or chanting as they marched. Ducil (Ducilius) the Irish Monk in 825 in Bohemia wrote the story of the raids by Vikings. The Culdees fled the Vikings from Norway, staying one jump ahead of them as the Vikings progressed to the Americas.

Driven from their monasteries by raiding Vikings, the Irish monks sought peace and quiet in Iceland according to Ari Thorgilsson, the Icelandic historian, who wrote in 1026. The question of how many Irish settlers were in Iceland and whether they left celibacy to interbreed with Viking or Inuit women intrigued many.

Benedikt Hallgrímsson *et al.* explored that issue through DNA. They published "Composition of the founding population of Iceland: Biological distance and morphological variation in early historic Atlantic Europe" in the *American Journal of Physical Anthropology* in 2003. They examined the founding population of Iceland through the study of the morphological traits in skeletons from Iceland, Ireland, Norway, and Greenland from the Settlement Period of Iceland. They found that Icelandic samples were much closer to the Norwegian samples than expected. They concluded that the Settlement Age population of Iceland was 60–90% of Norwegian origin and their findings did not suggest a significant contribution from Ireland or other sources.

The Irish, as a culture, left Iceland and went on to Greenland. They may have gone on to New England by the time the Norsemen arrived. Some believe that a site of North Salem excavated in 1937 in the Merrimac Valley in southern New Hampshire has Irish architecture.

About 1000 years ago, Iceland's storyteller, Ari Thorgilsson, wrote about Skraelings with a reference to Albania (suggesting albinos or white people) in the *Landnamabok*:

> *To the south of inhabited Greenland are wild and desert tracts and ice-covered mountains; then comes the land of the Skraelings, beyond this Markland, and then Vinland the Good. Next to this, and somewhat behind it (inland) lies Albania; that is to say, Hvitramannaland, Whitemansland, whither vessels formerly sailed from Ireland. It was here that several Irishmen and Icelanders recognized Ari Marson, the son of Mar and Katla of Reykjanes, whom there had not for a long time been tidings of, and whom the natives of the country had made their chief.*

The saga of Thorfinn Karlseffni who searched for Leif Erickson's house around 1007 tells of the capture of two Skraeling boys who spoke Irish. They gave Irish names, described their parents as Irish, and named two white Irish kings who ruled "a land across the water from them." Over time, these boys were taught to speak Norse. They finally said that their people had no houses but lived in caves or dens. They described their kings across the sea as Avalldamon and Valldidida. At the end of the saga of Karlseffni, it was concluded that the boys and their parents had come from Ireland the Great.

The Arabian geographer Al-Idrisi mentioned "Irlandah-al-Kabirah" or "Great Ireland" located "beyond Greenland" in his atlas of 1154 A.D. Norse sagas from the 11th century mentioned an Irish territory called "Ireland the Great" near the Norse Vinland colony in the vicinity of modern-day Massachusetts. This remarkable atlas of 1154 was the result of extraordinary collaboration between the Moroccans and Roger II, King of Sicily, according to Frances Gies.

During the mid-14th century, an anonymous Spanish Franciscan traveled throughout the New World and elsewhere. He wrote of his journey in a unique book known as *Libro del Conoscimiento*. The original title was *The Book of Knowledge of all the Kingdoms, Lands, and Lordships that are in the World, and the Arms and Devices of Each Land and Lordship, or of the Kings and Lords who Possess Them.* The *Book of Knowledge* was written around 1350 predating the Columbus voyages. It described all nations and their flags to aid Christian travelers, and described an Irish settlement in America without getting very specific as to its location.

Sir Clements Markham prepared an English translation of that friar's account in 1912. Here is an excerpt:

> *Being in Irlanda, I sailed in a ship bound for Spain and went with those on that ship on the high sea for so long that we arrived at an island called Eterns (Faroe), and another called Artania (Orkney), and another called Citilant (Shetland), and another called Ibernia (The Ireland of the North).*
>
> *All these islands are in the part where the sun sets in the month of June (the northwest Atlantic), and they are all peopled, well supplied, and with a good climate. In this island of Ibernia, there are trees, and the fruit that they bear are very fat birds. And the women are very beautiful though very simple. It is a land where there is not as much bread as you may want, but a great abundance of meat and milk.*

Well, all this is interesting to speculate, but whether the Irish were in this area of the Americas is only a guess. This journey description does not appear to have brought them to the west side of the Atlantic Ocean. They would have found little milk here, and America does have a shortage of trees growing fat birds.

Who else might have been near Rhode Island? There is a claim by some that the Chinese discovered the eastern coast of America around the year 1421. It is said that Zheng He of Nanking sailed with hundreds of ships to India, Africa, and possibly to France and Holland. There are also reports that the Basque sailors (who were fishing and whaling around Labrador in the 1500s) had sailed on to New England. The Basque people speak Euskara and have a language like no other. One day their origins might be learned because of a genetic anomaly: a very high rate of Rh-negative factor in the blood. Nevertheless, we're not sure that these people were ever near Rhode Island.

Christopher Columbus wrote his son, Fernando, that he visited Iceland in 1477 and even sailed one hundred leagues beyond it, discovering there an "unfrozen sea." He is said to have accompanied a Norwegian pilot named Jon Scolus, variously called Johannes Scolvus, John Skolp, and Jon Skolp. There is a notation that Skolp had accompanied Joao Vaz Corte-Real on a joint Danish-Portuguese expedition to Greenland the year before—1476. Morison doesn't believe that Columbus was on a voyage to Iceland with both Skolp/Scolus and Joao Vaz Corte-Real but others do.

We can be fairly certain that Columbus had heard of a western continent and had perhaps even seen artifacts from the Atlantic coasts. He then sailed on August

3, 1492, with three vessels, 120 men, and landed on October 12 on San Salvador. On three more voyages, he discovered Jamaica, Hispaniola, and Puerto Rico. Everybody wanted to sail to the New World and to find the Northwest Passage to Cathay after that.

However, it was quite some time before it was realized that what had been discovered was not Cathay at all but a new continent. Columbus believed that his new land was China, even insisting that Cuba must be Japan until nearly the end of his life.

His familiarity with the northern Atlantic seas is exemplified in his remarks about Flanders; Bristol, England; Galway, Ireland; Iceland itself and his visit even further to Greenland and to other Christian colonies in 1477. When compared with Marco Polo's description of Siberian Inuits and their peculiar dog sleds, the inescapable conclusion of the eastern Canadian/Greenland locale being the same as Polo's China/Siberia is logical. The Greenland Inuits were the same people; their dog sleds unique and identical to their unbelievably distant brothers several thousands of miles and maybe a full 90 degrees of longitude so much further west. The erroneous concept was real and it lasted.

Roger Williams stated that the Narragansett origins were "... from the north, from Tartaria." As well educated as he was, his outlook again indicated the long passage of time to today. While the Pacific Ocean had been discovered a century before Williams wrote this, news of a greater ocean than he knew and the concept had not yet reached much further than the Mediterranean seamen and scholars by 1643. Our task is to see the world as they saw it—no easy matter.

Modern science seems to have answered the question of presence and origins of the Greenland Inuits of so wide an expanse of the globe. It is my understanding that there were several immigrations of Inuits into Greenland, and that the most recent are the Thule Culture having followed the Dorset Culture. That transition occurred near the same time as Erick the Red. They did not have to be seafarers. But they were accomplished ice travelers with adequate cultural resources and skills to sustain themselves in long migrations over the entire roof of the world, given time. A mere twelve miles separated Canada from Greenland at Nares Straits, less than a day's walk over ice for an Inuit if the snow was not too deep.

There have been even wilder theories about how people might have gotten to Greenland and hence to America. A few have considered whether Siberians might have sailed across the North Pole Sea swept along with ice floes down to Greenland. That theory, called Nansen's Theory, was put to the test in 1893.

Fridtjof Nansen was a scientist, statesman, and Nobel Peace Prize laureate. His devotion to humanitarian causes saved the lives of countless thousands after WWI. However, he regarded himself first and foremost as an explorer and scientist. Earlier observations had convinced him that a strong east-west current must flow from Siberia toward the North Pole, and from there down to Greenland.

Determined to prove the truth of his theory, Nansen drew up the specifications for a ship strong enough to withstand the pressure of ice. The expedition left Christiania (now Oslo) in June 1893, with enough provisions for five years and fuel for eight years. Roald Amundsen considered going with him but refrained. Nansen's "Fram" sailed east along the northern shore of Siberia. About 100 miles short of the New Siberian Islands, Nansen changed course to due north. By September 20, at latitude 79 degrees, the "Fram" was locked in pack ice. Nansen and crewman Johansen prepared to drift westwards toward Greenland.

Conditions were far worse than expected. Their way was often barred by ice ridges or by patches of open water that caused delays. Finally, they decided to turn back and to make for Franz Josef Land. In August 1897, an expedition vessel spotted Nansen and crewman Johansen trudging across ice having lost or eaten most of their sled dogs. The men were rescued and deposited at the Norwegian port of Vardø.

The following month the unmanned "Fram" had shaken off the last of the pack ice near Spitsbergen. Numerous pieces of the ship were soon found at the southernmost coast of Greenland. Even though it seemed that Nansen's theory of a current existed, most people believe it is extremely unlikely that any prehistoric vessels and seamen could have arrived in America via the North Pole.

The discovery of the New England coast by the Icelanders is probably the earliest possible arrival of Europeans in that location. In the year 986 (eighty years before the conquest of England by William of Normandy), an Icelandic mariner named Bjarne Herjolfson, making for Greenland in his trading vessel, was swept across the Atlantic, and finally found himself in view of dry land. He made haste for a return voyage without landing anywhere, and succeeded in getting safely back to Greenland.

The story of Bjarne impressed an intelligent and adventurous young man, Leif Erickson. Leif bought Bjarne's ship, set sail for Vinland in the year 1000 with 35 men, and reached or explored as far as Cape Cod. He returned to Iceland in 1001, and two years later Leif Erickson's brother, Thorvald, established a colony back on American soil. I propose that it was at Narragansett Bay.

The nearest known populations of archaic Old World residents to the Americas were Scandinavian/Gael colonists and settlers of Iceland and Greenland. A critical factor for consideration of who could sail to the Americas is the nautical capability of voyagers over such long distances.

Seamen of Iceland and Greenland certainly did have sufficiently advanced and amply demonstrated nautical proficiencies sufficient to reach the area under study. Perhaps they were the only people of Atlantic environs until about 1350 A.D. with that kind of sailing capability. Pan-oceanic pioneering by their early maritime culture is reported from about 800 A.D.

The study of the literature of Iceland and Greenland documents four distinct expeditions and emigrations by more than 200 persons, male and female, from Iceland and Greenland to a district they called Vinland. These expeditions took place between about 986 A.D. and 1030 A.D., just about a thousand years ago.

Part VII

Additional Clues to Narragansett Descent from Vikings

Suspicion of a single, seemingly isolated, factor leads us to consider possible alternative indications of disease immunity.

Investigation and Commentary for Additional Clues to the Immunology Factor

In a previous chapter, I presented the argument that the appearance of an advanced level of tuberculosis immunity constitutes a scientific argument that Leif Erickson's Vinland has been recovered. I also showed that language similarities and certain other anomalies indicate something more than coincidental occurrences. Together with the dating of an incursion ("by an advanced culture") into Pettaquamscutt River Valley, this would seem to be conclusive. Additionally, there are indeed a number of factors that have been noted and documented by any number of historians without many noting or addressing their significance.

But this secondary discussion lacks the definite scientific provenance that might be wished. Many noted the general public health enjoyed by the Narragansetts when they transitioned from trading with the English, Dutch and French colonists to close and intimate dwelling with immigrant settlers in Rhode Island. This represents a period of over a century and a half without any notable deleterious effect upon their population. No other tribal group on either American continent, not even any Polynesian population, nor even the aborigine population of Australia achieved that level of public immunity.

Strange to say, this attribute is so well documented that many historians since that time have remarked upon it from varied viewpoints. The universal description of Narragansett "power" meant both martial and population density. Their so-called "golden age" meant prosperity at the transition era. Their numbers as a sizeable group have been recorded but the implications never examined, approached, or explained. In brief, it appears that Narragansett public health compared as equal and perhaps even superior to that of the incoming Caucasians. If we can rely on our historic sources, this is just as indicative of a general genetic resistance of European diseases as is the specific TB resistance.

We have anecdotal information from colonial records of both Roger Williams and John Winthrop the Younger. We also have the census of 1670 and the facts surrounding the Native American army attack of Providence in 1676. In the census, the native population of this area of less than 700 square miles could field an army of 5000 and did field one of some 1500 to 2000 in 1676. What this says of the population, which thereby must have exceeded 30,000, gives us a sense of strong resources in immunity to any and all diseases of Europe besides tuberculosis, which were primarily small pox, measles and others not precisely identified. The enormity of the terrible scourge of European diseases among the American aboriginal populations is little comprehended even today.

I was astonished at the wealth and integrity of sources who had applied themselves to this small but prominent Native American Tribe. Giovanni Verrazzano meticulously recorded his visit to the degree that we can easily see and nearly share the open and welcoming attitude of the natives to the visitors. One wonders at the thought that somewhere in Narragansett and Wampanoag American psyche was some dim, hidden memory of a time when they and their visitors had been more or less brothers upon the same continent. Verrazanno's visit was long enough and intimate enough in itself to have originated epidemics among the Narragansetts. But it seems to have had very little impact upon the populations as, if it did, recovery had been made within a century, which seems too rapid in comparisons with what came later to other Indian groups.

Roger Williams, whose sophistication and world outlook amazes us at a time when most of the pioneering settlers were constrained to simple agrarian lives and walking distance farm plots. John Winthrop, Jr. had training in law, diplomacy, astronomy and medicine. And yet we speak of a time so archaic that it is possible that none of these men were yet aware of the great Pacific Ocean, familiar to Spaniards, it is true, but its scope not yet into broader perspectives. The two latter described the faraway waters as "the Southern Sea" in the same perspective as the Chinese did with dense archipelagos numbering at least 7,444 islands.

Investigation and Commentary for Additional Clues to the Immunology Factor

This John Winthrop the younger (born on February 12, 1606, Groton, England, died April 5, 1676 in Boston) is often historically dominated and confused with his more famous Father (born in 1588, died in 1649) who was so long a guiding light to the Massachusetts colonists. Recent research on the Internet demonstrates the same embarrassment. This creates some confusion because the Junior's major American contributions were centered upon his Indian neighbors in Connecticut and Rhode Island—including our Narragansetts.

The painting of Indian Chieftain Ninigret was near certainly commissioned by this man and remained in his family for generations. Ninigret's painting is noted to have been executed by an "unknown artist" and so also is the painting of the younger Winthrop as depicted in *Encyclopedia Britannica*. It would surprise me not at all if the two exceptional paintings had been executed by the same person. The artist might have been a servant or a woman or maybe a felon from England reluctant to expose himself. It could even have been an illiterate Indian himself.

John Winthrop had followed to America a year after his father, who had sailed in 1630 to the Massachusetts colonies of Salem and Boston. At the son's arrival, he was elected "Assistant" (to his Father) and was active in establishing new settlements including that of what is now Agawam and Ipswich.

Ipswich is north of Boston and an expected satellite of the numerous new settlements near Boston Bay. Agawam, however, seems named from a highly regarded Indian Chieftain and the present day place of that name is well west and upon the Northern reaches of the Connecticut River. Still, some historians aver that modern Ipswich and archaic Agawam were one and the same place. At one time there was yet another almost adjacent to the Pawtucket Grant to Roger Williams, well south of Boston. This peculiarity demonstrates the extreme rapidity of expansion of the newcomers, as well as impotence of the declining Native populations.

INITIAL COLONIAL SETTLEMENT

At this time, virtually all of these pioneers were self reliant farmers living upon scattered plots often at some distance from each other where dawn to dusk hard work was the norm. People would gather in some central place, more than likely their church, for meetings and worship on Sabbath days, might visit one another from time to time, but generally were self sufficient upon their own properties. Occasionally a few of these farm plots were adjacent or nearly so for a small community but there were very few and very rudimentary "villages." The "general store," if it existed at all, would be more of a trading post where even Indians were welcomed. Roads and highways were rudimentary, primitive and generally followed existing Indian trails. The new system of American urban development as described by Mark Twain "… crossroads, then a church, then a liquor store, then a penitentiary …" was still years in the future. Indian populations outside of Rhode Island, if they existed in proximity to these farm "clusters," were failing and barely tolerated in the extreme religious prejudice of the time.

As far as I can detect, colonists seldom or never wrote about their Indian neighbors. It seems to me that this was from religious scruples when the clergy emphasized salvation and self worth only to the chosen—and the Indians were far from considered the "chosen." Apparently, there were quite a few converts to Christianity and these were referred to a "praying Indians" but even as such they confronted social strictures relegating them to less than equal status with the English.

There was an amount of social intercourse and probably there were numbers of Indian servants and farm workers among the English. At least one Indian had actually attended Harvard College. In Rhode Island, as I have attempted to show, the two disparate populations were on firm and peaceful accommodation. The Narragansetts often dressed as Englishmen, and entered both skilled and unskilled English trades. Their children were adopted into English families, where they were servants, blacksmiths, farriers and even a locksmith. A locksmith in those days was really an armorer, a high skill, the described "lock" being the firing mechanism of a musket.

By way of contrast, a few settlers were beginning to "go Indian," becoming so-called "Swamp Yankees." This, an overlooked activity sidelined to modern historian's footnotes, did not diminish and was a frequent source of embarrassment to certain of the more "elite" type of families. Benjamin Franklin, a century later, wrote that it was quite common and that of those who had once given the wilderness life style a trial, none ever returned to the English manner.

One of these, named Jonathan Tefft or Tift, had been present within the Narragansett stockade when it was attacked and overcome on a frigid December day in 1676. He was thought to have fired upon the English army, was captured and sentenced to being drawn and quartered and is said to have been the only recipient of this punishment in America. This macabre process was still prevalent back in England and remained so for many years. The English referred to the natives as "savages" and "barbarians," possibly because the Indians did not wear such colorful uniforms. From what I read, the colonial judiciary and military was easily as cruel and merciless as the Indian.

Apparently, the initial settlement of Providence was directly opposite and across narrow Moshassuck River and within hailing distance of a Narragansett village called Weybosset. (This is now the "downtown" district of the city and in particular the State's hub of its legal and banking professions.) Moshassuck River is not a large one, being just about one stage above a creek. Another river named Woonasquatucket of about the same size flows from the west into what had been a nearly landlocked cove. All three of these features at this point are tidal and in those days a prolific source of food right on the doorstep of the Providence settlers.

Another Indian village was one named "Mashpaug" (var. "Mashapaug"). This was a few miles inland but well within the scope of the boundaries of Providence that had been ceded to Roger Williams. (I knew this place well in my boyhood. There is a cemetery there now but adjacent to nearby Mashpaug pond was an area of swamp of unbelievably prolific wildlife with snakes, frogs, muskrat and at least one huge snapping turtle that was amazingly adept in eluding traps and ambuscades set by us boys. My experiences in and about this swamp came to the fore in my research, explaining why it was that Indians resided near and wintered in them. It was in these swampy places that life commences in early spring; early respite from the "hunger moon" of March and April. The swamp is now paved over and the wildlife gone to all but memory, alas, as are our childhoods. A fond memory of mine concerns a particular plant that arrives first in springtime in the marshes of Rhode Island.

Ode to Nature's offering—the Skunk Cabbage.

*Radiant jade amphora glowing, gleaming
green 'gainst winter's drabber duff.
Demure, demanding, illuminated herald trumpet.*

*Solitaire. Rising perfect at moon's command
to morning—and springtime.
Instantly, unfailingly reviving hearts and spirits
of winter's lonesome weary wanderers.*

The peculiar development of Rhode Island and Roger Williams so influenced by colonist's relationship with the Narragansetts might give added impetus to our considerations. My own outlook on it is offered here as, perhaps, a new perspective for other students with interests more specific than my own.

Providence Settlement

The crossing of the Seekonk River that was made by Williams is important to our story. Today, a ship approaching northward towards deep water port Providence is early aware of the sizeable river Seekonk bearing off east and near parallel with the short Providence River that lies straight ahead. But in colonial days this would have been obscured by a dominant hill of considerable size that was at one time called "Fort Hill."

This hill was the whole of a peninsula extending from the east shore nearly to the western shore, thus forming the first nautical constraint into the settled places. Erosion over the centuries has lowered the heights to near sea level and the sizeable geographic feature that it had once been has disappeared. Even the extensive peninsula upon which the hill once stood is now minuscule.

In my boyhood, I visited the place a number of times. At that time, there was a modern stonework replica "fort." It had replaced the olden one that existed in what is now empty space a quarter mile west, overlooking the Providence channel. I would judge the height of the promontory from old illustrations at perhaps 350 feet altitude. The whole of it, hillside and fort together, have now gone to the elements. But even when I first came there, it still offered a spectacular panorama in all directions. That limited height has again lowered by more erosion removal of at least 300 horizontal feet.

I realize now that at one time there must have been a group of Narragansetts, Wampanoags and pioneers including Roger Williams that gathered upon this forgotten hill. This had been the only place from where all the boundaries of

Roger Williams' grant and also the placement of the English settlement could have been seen from one spot.

From the promontory at that time, one could also observe close by the northward tending river bank upon which Providence was originally settled. These grant boundaries were defined from the just visible Pawtucket Falls (Blackstone River) to the north, the obscured Pawtuxet River Falls to the south and to the visible but distant Neutaconcanucket hills some eight miles west at a place called "Hippses Rock."

These delineations framed an immense area even larger than the present day boundaries of the city. I am assuming that famed Chief Massasoit had been present in this group, as he had made, or was to make, the Pawtucket grant to Williams that extended an even larger area north of the described "Pawtucket Falls."

In combination, these grants, besides that of Prudence Island (not small) and land for a trading post at Wickford, cover a huge area. If we take Roger Williams' word for the density of Narragansett villages, there must have been numerous Indian towns, including Mashpaug, more than a few minor and major chieftains, and sizeable native populations within the scope of this terrain.

It has always been a mystery to me why the Narragansett and Wampanoag Chieftains were so enthralled by this man as to make these gifts to him. As I have suggested, they might have been appreciative of Williams' outspoken belief that lands seized by the colonies from the natives should be compensated.

There are hints, but apparently no surviving recordings, of Narragansett presence at certain events in England's developing revolution and knowledge of Williams' rising faction in British politics. Williams had been closely involved in the recent war upon the hated Pequots, the victory of which was mainly to the Narragansetts. These grants might well have been in the nature of rewards for this. Williams stated that his purchase of Providence was for thirty pounds (sterling), but Canonicus, the grantee, claimed it was for "love alone" that he gave this away.

It seems politically naïve to assume that he "gave" it away, but there were enormous differences in how settlers and Indians viewed these "purchases." The English attitude included the concept of "trespass," the right to exclude or evict people who did not own or paid no rent for the land. But the Indian felt that no one "owned" the land and that only possession, family and tribal alliances, and military ability were the defining factors.

When they received what the colonists believed were payments for the properties, the Indians viewed it otherwise, feeling that they were simply accepting gratuities for friendship, mutual tolerance, or permission to utilize certain areas. We are all familiar with the sale of Manhattan Island for 24 dollars in trinkets and other negotiations of the same sort. Massasoit's deed to Aquidneck Island (Newport/Portsmouth) included a proviso for "four coats" which happened to be delivered in the wrong size or color but the "sale" was allowed anyway.

We now look on these negotiations in the same light as the colonists of Viking Thorfinn Karlseffni had in his earlier trading—that the "skraelings" were naïve, ignorant of value and therefore somewhat inhuman and subject to, even deserving of, easy exploitation. This was not so; simply a different set of values and goals in their transactions. We certainly cannot enjoin the Indians for venality, and who can truly say who owns the land on which we stand? Where is the value? Sometimes useful earth, or the people who make their lives upon it?

However, it was many years before Providence grew to any size. It consisted of only some 30 or 40 dwellings arranged along a narrow level riverbank facing west both river and cove and seems to have remained near this size and configuration for at least 50 years. It seems logical that some new settlers came in over the years and settled at varied places away from but near the new colony.

Seventy years later, in 1717, it was still a "backwater," not even identified on British Naval charts. The manner in which these properties were laid out along what is now North and South Main Streets is curious enough to give grounds for some thought. The properties on a map show them as rather long and narrow, being maybe 150 or 200 feet wide but quite long in an eastward direction, perhaps three quarters of a mile. They were on the east side of the river only, on the eastern side of the road and extended to what apparently was a busy Indian trail which is now Hope Street.

At first, this would not seem too unusual as this *size* of farm became typical in later years. But what the ancient maps do not show is that the dwellings at riverside abutted right in their back yards a steep hill for a good half of their area. This hill is so steep that it had little value for growing crops. If it had ever been plowed, it would have sustained the same extreme erosion effects that had occurred just south at Fort Hill. The hillside could have been used as pastureland, cattle not being over-particular. Unfortunate settlers must have been constrained to long daily climbs to where they could, doubtless winded, plant, maintain and harvest crops. Brown University campus and properties now occupy a considerable portion of those ancient hilltop fields.

These plots were termed "plantations." They contribute to the State's history and lengthy name of "State of Rhode Island and Providence Plantations." "Rhode" was Aquidneck Island where the Newport and Portsmouth Colonies existed. Providence was where the followers of Roger Williams dwelt. The entire state otherwise was "Indian Territory" except for another settlement of Warwick which was just south of Providence's southern boundary of Pawtuxet Falls.

As usual, my boyhood contributed a bit to this as well. I recall locating a small, overgrown and isolated cemetery in Warwick that had on one headstone a death date and age of a man who had held the rank of Captain, probably sea captain, that by transposition discovered his birth to have been in the late 1600s. I wish I could recall the name, but at the time, I felt the headstone and his memory would last forever. I think the death date was somewhere near 1740. A house was later built over the small cemetery and I often wonder if the inhabitants heard strange noises or thought of the interesting history to which they were close neighbors.

The matter of such close proximity of English Providence and Indian Weybosset as being within actual earshot has never been addressed by historians, so far as I know. Ancient maps of Providence do not show any detail of the huge grant whatsoever except for trails outside of the town itself. Yet, Weybosset as an Indian village must have existed there, for it is inconceivable for it not to have been.

Narragansetts were noted for waterside dwelling and the area at the terminus of what is now Weybosset Street was ideal for Indian settlement. For one, this confluence of three rivers entering salt water made it a prolific source of seafood. The rivers and now absent cove were replete with all manner of clams (soft shell), quahogs (hard shell clams), mussels and many seasonal fish.

Few Rhode Islanders today can imagine that even into my boyhood in the 1930s this upper bay and harbor of Providence supported major fisheries of scallops and oysters. Roger Williams' crossing at Seekonk is narrower than in his time largely from tons and tons of discarded oyster shells that had been extracted from the waters nearby.

These conditions of prevalent protein food resources right at their doorsteps must have been a monumental and critical asset to the early settlers and attractive; vital to both pioneer and Indian alike. It strains credulity that the Narragansett Indians would depart such a favorable place, given their close bonds with the early residents of Providence.

Intimacy of the Races

The purpose of this examination of local history is to determine just how intimate the Narragansett Indians were to the colonial settlers as it might effect disease interchange. Most historians of Rhode Island and nationwide would greet my suggestion of this dwelling mode occurring simply across a shallow river with surprise and disbelief. It is presented here as "suggestion," since I cannot yet prove it. It is much more than suggestion and comes from tracing the "—sett" linguistic suffix into tangible locales throughout the Narragansett Bay environs.

I knew many of these places well and it came as a shock to discover that they had once been residences of dominant Indian chieftains living in close approximation to the lifestyles of ancient Viking "Jarles." I discovered Weybosset early on as I had traveled that Providence street many times and could well find my way anywhere along it in my sleep. It was only a short step to conclude that this had once been yet another waterside village of Narragansett Indians as it surely must have been.

When Roger Williams encountered the five sentinels guarding the Seekonk River crossing, it is more than likely that these five were residents of Weybosset, less than a mile distant from the guard-post. In my mind's eye I can see them directing Williams' small boat party around what is now Fox Point to, not his spring, but to the chieftain of Weybosset where initial negotiations commenced. The two already knew each other because Williams had been trying to farm a small nearby plot for some months in the territory that Massachusetts laid claim to and from which Williams was escaping.

I was even more surprised to see upon an ancient map the location of the Indian village Mashpaug, a common site of my boyhood excursions. This brought me to the realization that these villages were not always "somewhere else" but right before my eyes, if only I could see into time itself.

So there was a Weybosset and it must have been a waterside location. The area had been a prolific fishing ground and I suspect that even salmon came up this far into the then pure waters of what is now Providence harbor. The friendly relations with Roger Williams are evident and the grant soon, if not immediately, concluded. I can see also an adjournment back across the River to nearby Fort Hill, which then dominated Weybosset, the future Providence, the Seekonk crossing, and all the borders of the Providence grant.

Williams was not under "hot pursuit," so one can imagine these negotiations taking place with simple sweeps of the arms and points of the hands. The fort which came to be built much later, was surely a relic of the Revolutionary War when Newport had been occupied by British troops and the lower bay dominated by the British Fleet. These circumstances, as might well be imagined, were discomfiting to Providence residents and merchants. They felt it necessary to defend from possible British assault on their city. Thus they constructed two forts, this one on Fort Hill and another further south on the west side of the bay near what is now Thurbers Avenue.

I visited both and consider them as ideal placements for defensive works for cannon. The one upon which I claim these negotiations took place seems to me to be more than ideal as one would not need much of a cannon to reach approaching ships. One might almost drop cannonballs down upon their decks from where it seems the fort had existed.

We gain insight into the intimacy of the Narragansetts with the slowly growing population of English settlers in Rhode Island. Many historians overlook the century and a half from first intimate contact and fifty years of amicable co-existence of these disparate peoples. In their view, we might expect the local natives to have experienced the rapid decline from epidemic diseases that all others of the Western Hemisphere did suffer. First, we have to consider that this catastrophe did not here occur. Then we must wonder at the miracle: and then we find that there was, in fact, a reason and a cause. This cause is another explanation of an earlier, near unimaginable miracle of an adventurous journey down an extensive and unknown coastline.

In his historic letter to his brother in April following the 1676 attack, Roger Williams reported that he reminded Narragansett chieftains "... *this house of mine now burning before mine eyes hath lodged kindly some thousands of you ...*"

This home was within the scope of the new town itself and the remark speaks volumes to the relationships and intimacy of the natives and the colonists. Note well the numbers he mentions. Imagine just where he stood when he spoke to the chieftains whose huge army stood by. It was very nearly within the bounds of the town itself. In fact, we can place it rather closely. It was either upon the spot now occupied by the white marble State House, or somewhere upon its grounds.

This area overlooks the Roger Williams Museum and his "spring," so close that people might communicate in loud voices between the places. The home that was burning, however, while nearby, was not at that precise site. Just as a matter of interest, the "thousands" number that he actually used in his face to face encounter would have been "Nquittemittannug," in Narragansett, one thousand.

We see here through the mists of time the poorly recorded perspective of Indians and colonists in the long period of peace strolling the same streets in close proximity. This was true also in Plymouth where Indians, even Narragansetts, were frequent visitors and often conferees. In Rhode Island, we are not dealing with hostile, disparate and isolated populations, but in fact neighbors with differing mores who yet had ample opportunity to contract each others diseases. Elsewhere and everywhere, these conditions were mortal to the Indian. For fifty years, these conditions prevailed in Rhode Island. And yet, this Native American tribe abided in prosperity and constant numbers, enjoying similar survival rates to that the colonists themselves recorded.

We Hear Again from John Winthrop, Jr.

John Winthrop the Younger received his education at Trinity College in Dublin, Ireland, and followed this up with two more years of law. He later "went to sea" and traveled to France, Holland, Italy and even Lebanon. He found time somewhere (possibly at sea) to also study medicine and was considered a physician at the time he moved in 1635 to Connecticut, where he treated, on average, 12 patients a day. He was active in mining and developing local mines in New England and also extraction of salt from seawater. He also owned an astronomical telescope either of 8-inch lens diameter or variously three and a half feet long. He apparently was the first discoverer, but not credited, as observing the fifth moon of Jupiter. He participated in the seizure of New Amsterdam from the Dutch in 1664, after which it was renamed New York.

Here we see an educated and enlightened man of acute observations if ever there was one. He was the very next door neighbor of that segment of Narragansetts called "Niantics" and a close friend and military ally of Ninigret. While Williams was so close as to actually dwell in their homes and travel with them in parties armed for war, the relationship of Winthrop the Younger, diplomat to the core, may not have been that intimate. Winthrop was certainly close enough to know them well and know an epidemic if and when he saw one. And in this case he did not see one where he expected to—and said so.

One is reminded yet again of famed Sherlock Holmes' interest in the dog that did not bark in the night time. To paraphrase, Narragansetts did not die in the pestilential time. As with Holmes and Arthur Conan Doyle, this literary detective work seeks to identify the culprit that was responsible for this curiosity.

Now, the Connecticut colonies are subject to some confusion because of the distances between major settlements as well as having close relationships with the Massachusetts Colonies. Near New York/Amsterdam was the burgeoning colony of what is now New Haven, a seacoast entity. Miles further east lies the large Connecticut River whose delta was settled at a place called Saybrook, and this is where the younger John Winthrop came first to settle—by contract with one "Lord Saye" in 1635.

The Indian tribe in this area was the Mohegans but the new English settlement did not prosper. This river is navigable for many miles inland (north) and was a major route for Indian trading (furs) and was being settled by Englishmen some fifty miles from the sea. Eastward again is the area of Mystic and New London whose local aborigines were Pequots. New London is also upon a river, but this one does not have an extensive run into the inland hills, being more or less a narrow salt estuary or perhaps a small fjord. Both Connecticut and Thames Rivers are very hilly even at the seacoast, perhaps even so far as to be called mountainous. Their abrupt and lofty riverbanks are most impressive in autumn when the foliage glows brightly in magnificent color, so delightful to the eye as to defy description.

To confuse the issue even further, the 1636 war upon the Pequots, unpopular people with just about everyone, was instigated by the Massachusetts Colony. Both father and son Winthrop as well as Roger Williams were instrumental in initiating this attack. The relationships of these men seem at first to hint that it might have been an intent to surround the Rhode Island territory, thus encompassing Narragansett Bay into Massachusetts. The Narragansett Tribe also participated as mercenaries of the colonists with the conniving influence of all three English parties.

To their credit, the Narragansetts had second thoughts after their victory by the apparent genocidal policy of the English troops. They complained at the massacre of women and children, which was a new thing anywhere in aboriginal native warfare. Once this perspective is taken into account, it becomes apparent that the present border lines of Rhode Island were drawn, not in England by political fiat as all other colonies were, but by the presence of the still powerful Narragansett army and populace.

The peculiar situation of the Rhode Island colonists, being predominately exiles, criminals and religious heretics, together with the tolerant and welcoming attitudes of the Narragansetts made the two intermixed neighbors natural allies.

At this time both populaces must have felt that they were one people with neither having any particular affection for the Massachusetts and Connecticut colonies. Quaker presence in Rhode Island (members of this sect were sometimes hung in Massachusetts) was influential and probably more than a little responsible for peaceful coexistence until 1676.

Colonial lawyers had forced the surviving Indians into indenture and slavery as reparation for King Phillip's War—they had little but their bodies and their labor for repayment. Some repayment was made in their own coin of Wampum(peague), but that medium soon disappeared from the colonial economy. Many Narragansetts were transported to the West Indies as slaves where they mostly died off in the hideous working conditions of the sugar plantations there.

At the time of the birth (c.1757) of founding father Alexander Hamilton in Nevis, it was said that of all those slaves brought to English West Indies from Africa, four out of five Negroes died within five years from overwork, disease and neglect. And the Narragansett survivors brought there had not even that "success" rate. It was, after all, only a year or two from when they had been free people with their own territory in which they could move about at will. Removal was a death sentence for them, in fact. Those allowed to remain in Rhode Island contributed to that strange statistic that their area at this time became the most concentrated population of slaves (51 percent) ever in North America.

Rhode Island ship owners had been early and earnest in the American slave trade. Old family fortunes were made in this despicable business. With absolute impartiality, the slavers carried both Indians out and Negroes in, but Rhode Islanders otherwise seem to have made early endeavors to both alleviate and eliminate slavery as did the other New England States.

Narragansett Indians Decline

Narragansett decline was precipitated after 1676 and there is little indication that disease had much to do with it. King Phillip's War ended with the collapse of Indian logistics. They could not plant and their numbers were decimated in surrounding Rhode Island. Already scattered, Narragansetts adopted the typical philosophy of the defeated—every man for himself.

Roger Williams claimed they could run through the forest eighty to a hundred miles in a day, so many availed themselves of this skill and took up abode with other tribes at considerable distances. Some Indians of faraway Wisconsin claim Narragansett blood. Williams reported that even at that early period, Narragansetts had become skilled equestrians and with these horses used as pack animals, they might move bag and baggage to distant removes. What was left of the tribe remained in place where their descendents still live disbursed among the general population. Their cultural center is a small reservation in Charlestown, Rhode Island. Some look "Indian" but many are indistinguishable from the average American of today.

Eventually a new organization called "New England Confederation" was formed in Boston to encourage cooperation and consolidation of all the disjointed New England colonies. This organization soon ruled that the Connecticut Colonies were separate from the Massachusetts Colonies. Rhode Island was still essentially considered a Narragansett/maverick exile territory separate from the rest. It does not seem that it was an early part of this confederacy of New

England colonies. Amazingly, the original charter for the Connecticut Colony granted it territory all the way to the "South Sea" (Pacific Ocean)! Indeed, a Rhode Island settler later made a claim that the borders of the Pawtucket grant north of Providence also extended to that distant ocean. (This came from somewhat vague descriptions of where the western border might be, along with Narragansett political naivete in acceding to a request that colonial cattle might graze to any distance in that direction. The ambitious pioneer claimant apparently considered the cattle free to walk, and he to sell, all that land three thousand miles to the Pacific.)

It might be noted here that this issue of cattle grazing had considerable import in early pioneer/Indian relations. Indian fields were not fenced and could be defended from wildlife encroachment by scarecrows, dogs, and Indian children. But the Indian was not then familiar with the handling of the large bovine, hog, and horse of the English and could not control entry into their fields. They had had the same problem with the raging bull owned by Thorfinn Karlseffni many years earlier.

Consequently, when these animals escaped their confines, a frequent occurrence long resolvable in law among the settlers, it could become a major disaster for the Indians. A prime factor in the commencement of the Pequot War was the issue of Indian claims for damaged fields against both settlers and their governments, all of whom felt the matter more conveniently resolved by annihilation of the Indian. The deed for the sale of Aquidneck Island (Newport) negotiated by Massasoit with Rhode Island settlers specified that Indians could remain on their land but, if so, must fence or depart.

In 1640, as a sign of respect for their previous Governor, the Connecticut Colony granted Fisher's Island to John Winthrop, Jr. and here, in 1646, he enters our story. For Fisher's Island lies at the confluence of New London's Thames River, Pawcatuck River and the present border of Connecticut and Rhode Island. Here he became the next door neighbor of famed Ninigret. It is at this time immediately after the "Pequot War" that I am certain the painting of the famed Indian chief as well as of himself were commissioned.

While Winthrop Junior died in Boston in 1676, he had only been visiting there for a meeting. The latter half of his life he had dwelt at Fisher's Island and New London. Fisher's Island has many distinctions; appearing partially at center western top margin in our peculiar, oddly oriented perspective of the New England coastline.

It is the contention of this book that it was the very island that had been visited by Thorvald Erickson some 640 years prior to Winthrop's residency. It is quite close to and within visibility of both Connecticut and Rhode Island (an ideal destination for a small "afterboat"), but despite this is at present administered by the State of New York, from anywhere in that State invisible.

It is not known who the native population of Fisher's Island were, whether Pequot or Narragansett, but it was likely an area of contention between two hostile tribes and possibly either uninhabited or sparsely so. When Thorvald Erickson landed there it seems to have been uninhabited but with some sort of agriculture by short term visitors. This is the same cultural attribute so well exploited by Vikings—military weakness of island populations.

And it is John Winthrop Jr. who joined with his friend Roger Williams in a definitive statement that the Narragansett Tribe did not suffer the pestilences ("plague") that even then were destroying so many other Indian Tribes. Their word "plague" meant, as I see it, the general decline from all introduced diseases, including tuberculosis, that was crippling all other Indian groups. I hesitate to use the word "decimate," for it was much worse than ten percent dying but more in the order of not even ten percent surviving.

The statements of these two and other historians can best be described as "anecdotal" but are supported beyond that to "circumstantial" by the more definite recordings of the census of the tribe in 1670 and the size (1500 to 2000 warriors) of the army that attacked Providence in 1676. In my opinion, this factor then is arguably close to certain.

In all this catastrophe that overtook the American and South Sea populations, all declined, many were extinguished, entire peoples disappeared from existence by reason solely of introduced diseases. And of all these peoples, all those many millions of persons, all those hundreds and thousands of groupings, one, and only one, people proved the exception—Narragansett Tribe of Rhode Island, whom we now justifiably aver descended in part from Norse Europeans with their immunities.

Some Notes from Professor McNeill

I turn now to a masterful preface in William H. McNeill's *Plagues and Peoples* where he develops in an exquisitely literary manner how his interest and work was based on certain remarks in Colonial Spanish records. He wondered at the phenomenon of such small numbers of conquistadors managing to overwhelm myriad populations of Native Mexican tribes. The factor transcends the arguments of political sophistication permitting alliances with them, Native awe of horses and superior Spanish armor. Indian fear of equines disappeared with the first observed deaths of the horses and Mexican armor was not that inferior to Spanish. There are some records of Spaniards actually using certain items of Indian armor, particularly the fabric body armor so much more comfortable in heat and movement than Spanish iron.

It is now generally accepted that disease was a major factor in the takeover by the Spanish. Disease alone devastated but its accompanying psychological demoralization was an added burden to the unfortunate Indians. Records show that even the general belief of the natives was that the pestilences were curses of the gods. Indians in Mexico and everywhere else felt and believed that they were abandoned by their established gods. If so doomed, any resistance seemed futile. They were well aware of their decline and this belief in doom resulted in a common attitude of resignation and defeat.

Demoralization commenced so early that when most pioneer settlers first described them they were often beggars, thieves and drunkards, nowhere near their lost manhood—the result of decay of their lost stable and constant cultures. A notable example of this attitude is that of President Abraham Lincoln so well noted otherwise for his sense of humanity. He had fought briefly against an Indian insurrection in the Black Hawk War in western Illinois. His attitude towards Indians was somewhat less than would be expected, and his government's policy toward Indian resistance was much less than what we would expect of a humane person. Lincoln had never known the Indian in his noble state and forgotten culture, only the desolate, despairing and often alcoholic survivors begging near the general stores.

It might be noted here that alcohol itself is an introduced toxin following the same rules of evolution for survival. "Drinking" societies create a segment of people who can "handle their liquor" and survive exactly as they would to a disease. Individuals, especially in primitive societies, often fail. This is a syndrome still observable among Native Americans. Here in Arizona with its 22 Indian reservations the problem of alcoholism is still prevalent. Most Indians who succeed in entering the urban societies solve the problem by simply not drinking. Their youth often come into the cities in curiosity and often collapse after drinking but one can of beer. One might compare with certain of European societies where drinking is prevalent and often a test of "belonging." The Russian ritual of non-stop toasting till all are under the table comes to mind, but other cultures also practice somewhat the same.

Professor McNeill discovered that even the time of Cortez' defeat and hairs-breadth escape from Tenochtitlan (Mexico City) coincided with an actual epidemic of small pox raging within the city. The Mexican General who first defeated Cortez had himself fallen victim to it. Incas of Peru suffered the same fate once Pizarro appeared. From this time onward, Mexican and other Indian populations adjacent to Spanish incursion declined to very small numbers. McNeill describes the effect as similar to what is observed in a petri dish in the presence of hostile (antibiotic) microbes. Resident populations of the tiny organisms appear to shy away from the intruders. This, of course, in the petri dish is not a result of conscious avoidance but of outright mortality in proximity to an incoming invader. Each population produces its own unique environment, which may be hostile to another population. And so was the tragic result with the aborigine Americans.

Plagues and Peoples deals with factors that influence the spread of diseases and how history, population density and land area affects immunities—something in the order of a "law of unintended consequences." The professor describes variances in disease and immunity from the growth of human use of wool and relationships between sheep, humans, and species of rats.

McNeill describes such informative factors as dwelling styles of two species of rat that influenced Bubonic Plague. *Pasteurella Pestis* enjoyed a strong affinity to the flea that infests black rats but a less strong one to Norway gray rats. The black is a good climber and often lived in the common thatch roofs of medieval dwellings. Rats, feces, and fleas could fall to floor, seat, and table, which concealed all manner of microbes. Gray rats prefer outdoor burrows, enter dwellings in search of food, and therefore are not such intimate neighbors as the black rats. These two factors (invasion and dominance of gray rats over the blacks and the use of tile roofs instead of thatch) reduced black rat populations and alleviated Bubonic Plague. From such simple things come enormous consequences.

While Professor McNeill's analyses and discussions sometimes become incredibly complex, even while remaining quite plausible, the discussion concerning the premise of this book is rather simple.

We see "natural selection" at work. Sensitive and naïve individuals succumb, while "exposed" or "experienced" individuals survive and become more likely to reproduce themselves. Those surviving have some ability to resist or overcome a natural enemy. They reproduce themselves with progeny inheriting the higher "odds" to resist or overcome any enemy, seen or unseen.

Clearly, the Narragansett Tribe that was encountered in 1524 by Giovanni Verrazzano, was in many ways unique. From this we can conclude that there were both cultural and genetic factors at work that enabled this, even if the genetic factor might be so complex that the science of DNA analysis might fail us. However, it does appear that in the case of certain disease immunities, some progress has been made in locating genetic anomalies upon DNA strands that seem to indicate immunities. It is my belief, however, that the resistance that signifies immunity might be of such complexity of numerous factors that we may be constrained to observe it by observation and statistical analysis alone.

Science fiction often delves into tales of ancient bodies (human, animal, reptile and even dinosaur) coming to life after centuries of sleep or burial. This is also the theme and hope of those trying to develop cryogenic preservation of deceased persons. But we can see that any life that can be re-constituted might well be doomed because it had removed itself long enough from the dynamics of natural evolution.

Rip van Winkle awakened after some 20 years of sleep and his fictional story assumes that he survived for some time. The reality is more than likely that in some brief interval he would encounter some lethal microbe that his neighbors had unknowingly fought to a draw, but which in "old Rip" would be child's play to some simple and malicious single cell organism.

His experience would most likely resemble that of Ishi, last North American "wild" Indian (of California). Ishi attempted (by surrender in 1906) to enter the world of the Caucasian. His family, his tribe, his entire people had died of disease. No other person in all the world spoke the language of his birth. Ishi had been kept isolated after his surrender for about five years by compassionate Americans and in that period had prospered. He expressed a desire to visit San Francisco. His saviors, who had made every attempt to protect him from disease, saw no harm in a short visit. His stay there was only a few weeks or a month, but long enough to contract tuberculosis and he died within two years. Thus was the end of the wild North American Aborigine.

To our (we invaders) credit, after centuries of malice and neglect, this man was supported and aided in life, grieved and mourned in death by some of us, the new people of the New World. We are not all of central European descent and admixture. The modern Hispanic of Mexico and South America is a brother culture of the Narragansetts, for a very great many are the result of intermarriage among the Conquistadors and Indian populations that grew up about the Catholic missions.

Time is a crucial part of this argument. Initially we see the Mongolian forebears of the Amerind removing themselves from the microbial pool of the Old World, becoming isolated by stricture of the Bering Straits, and commencing a separate line of evolution of disease and immunities. The result of this was the eventual conflict post 1492 when the Old World syndrome dominated the New by having evolved in a larger and more dynamic microbial pool of Europe, Africa and Asia.

The sparsely populated and more isolated American populations did not develop and did not even at that time of Bering Strait journeys possess the all too familiar small pox microbe. Small pox was a relatively new disease indicated by its "florid" symptoms and volatility. McNeill theorized that small pox was the disease, new to the Mediterranean, that invaded and spread rapidly throughout the Roman Empire in 165AD, although it might have originated in India. It seems now to have been eradicated world wide from a concentrated effort by United Nations medical teams in mid last century. When last seen, however, it was still quite volatile with its unique florid symptoms, and quite capable of initiating and sustaining widespread epidemics.

Another time conflict came with the invasion of English setters in New England. It concerns the issue of how long it takes a population to survive the "florid" first assaults of disease, and increase their population to previous levels. This time differential must be addressed. Recent history shows that even from 1492 very few, if any, native populations have even come close to a recovery from most European diseases, although some have made progress with modern medicine. Reason demonstrates and Professor McNeill records that it takes many generations and many cycles of endemic/epidemic experiences to accomplish this. Narragansetts, by all appearances and recordings did not have the time required from 1492 to fulfill this time line, and yet prospered. The cycles, still far from concluded, for TB remissions must have taken centuries, and these centuries must have been at some other place at some more archaic remove from where we discovered them in 1524. The reader is again reminded that these people as presented by this author are not entirely direct descendents of Vikings but a new combination of people by intermarriage of Natives and those Norsemen of Vinland who preferred to remain rather than return to the Northlands.

Our Earth turns and we with it. Those who would stop the world for the opportunity to get off, have no reasonable destination. Worse, should they change their minds, they will risk death if they rejoin after some interval. Even short residences in space stations seem to have some effect in astronauts in those aspects of human health over and beyond the weakening of bones from lack of gravitation. We are creatures of progress and the evolution of time whether we would or no. We live in both space and time. This world, this Earth, this speck speeding unimaginably quickly past a specific place in the cosmos is ours and ours alone. And this *time*, of all that infinite to the past and infinite to the future, is that in which we live. We perpetually increase in population as a result of temporary control of disease and transformation of the environment which we inherit. But we risk unknown and unimaginable disasters that await unless we tend carefully what remains of the garden that is the legacy in which we live. Our microbial enemy is unseen, elusive, and effective. Mother Nature impartially provides for the simple and minute as well as to the complex and exalted those same implacable adaptations to *time* that we ourselves possess.

Part VIII
▼
"Refugio?"

In which we delve into local colonial history for additional information. We find much of incidental interest, some of which is in support of our theme. We compare the Viking landings to those of certain original "Yankee" arrivals. We find them closely comparable and we end with commentary concerning two who are certainly the most perfect resources available to scholars of this subject.

Was Jireh Bull's Property a Viking Site?

Let us now examine property close to, but invisible from, the sea. It is difficult to approach, rocky, steep, and seems to modern eyes an unlikely home site. Yet, from a seaman's standpoint, it is a natural and favorable place for an explorer or settler to make residence.

Narrow River mouth is a difficult approach and would seem to entail a grounding by an unfamiliar seaman. Nevertheless, this is what the sagas actually say about it. Once in the river and up to the lake, the water is shallow with several clearly seen (from a masthead) channels that necessitate where the ship might travel. The ship must turn to the north and in that direction appears Pettaquamscutt River flowing between steep banks into the lake. To the east, the water is shallow. The shoreline to the west has a near approach to a low shoreline consisting of a sort of peat made up of sedge grass, a site that is ideal for landing. Indeed, there are "notches" in the sedge there that evidence that at one time three ships moored there. Sedge (peat) is a living organism and resistant to erosion because of constant re-growth. Whether those three notches could be a thousand years old, I cannot say. Very likely, they are the remnants of the shipyard that existed here in colonial times, but it is also true that the fleet of Karlseffni consisted of three ships.

The flat marsh of this sedge forms a "point" in to the lake. Just to the north on this side is dry land suitable for residences (as it is today) or a shipyard, as it was near 1660.

Looming above this area is a high, steep hill, which has always been forested, and from this, we might compare again the description of Hop that "... there were houses down near the water, and more farms further back over the hill." This is precisely the arrangement that exists today.

Two men made this approach from the viewpoint of first time entry, Leif Erickson and a man from Newport, seven miles across the Bay, named Jireh Bull. When Leif entered this lake, his saga seems to say that there were no local residents and this is confirmed in archaeological records of the place. When Jireh Bull entered 660 years later, it was Indian Territory but recently purchased from Narragansett Indians. Leif's goals were to camp, own, and hold for something in the future; Bull's goals were to set up a shipyard and residence.

Despite the 660-year difference, these two men had in common similar culture, social structure, and even ship design. This being the case, we might be able to make presumptions as to how and where they would found their settlements. English social structure nearly always had dominant social families residing on hills overlooking areas of commoners dwelling below. Norwegian culture was similar even if dictated by hilly fjords.

A truism of seamen is that a ship owner or captain will be always alert to his ship. A ship is never safe, either at home or abroad. It is constantly absorbing materials and work of its crew. A ship owner will always reside near his mooring and, if possible, within sight of it.

Therefore, now that we have found the likely landing and plausible residence sites, we may look up the hill for more information. And lo! We have discovered that the home site of Jireh Bull was, indeed, located up there overlooking both the shipyard and homes of the 17 people who made up the shipwrights. It is a steep climb but apparently not a discouraging one. Can this give us insight as to where Leif Erickson built and where Freydis dwelt after evicting her partners? Might we now look for evidence of the home of Karlseffni that the sagas say was south of it? Can we look north for the cliffs that the sagas say were near the river and a place of refuge from attacking and pursuing natives?

Following this line of investigation, I was able to confirm the many factors described in the sagas. Not only do all the landmarks exist, but also it is possible to explain the battle that took place there and compare it with the later battle of 1676. Finally, in the 1917 excavation, there appeared traces of structures not connected to the Jireh Bull home site that had been built in 1663.

Rhode Islander Jireh (Jireth) Bull was the son of Elizabeth and Henry Bull. Henry's garrison house in Newport will be described below. Henry Bull was a British immigrant who settled in Newport and after serving as a sheriff would become governor of Rhode Island from May 1685–1686 and again in 1689–1690. He was a Quaker follower of Anne Hutchinson who was banished from Massachusetts and settled a colony at Aquidneck.

Henry had arrived in the Massachusetts Bay Colony of Sandwich in 1635, but moved to Newport with some other men. On June 4, 1638, by common consent of the proprietors, they set aside 300 acres of land to support some minister (who would be Roger Williams). Others soon joined these seven men, the founders of the new country around Newport. Within ten years, the whole tract from Point Judith to Wickford (15 miles) was being purchased for projected development, but by no means heavily or with dense populations. There were probably no more than 50–60 English settlers dwelling on that side of the bay before 1676.

Jireh was born in September of 1638, just after his parents settled into Newport. He grew up on his father's property, which abutted that of Benedict Arnold according to Arnold's 1677 will that stated:

> *I do give and bequeath unto my dear, loving and beloved wife, Damaris Arnold ... ye house and two acres of land, be it more or less, that I bought of William Haviland in ye precincts of ye town of Newport, ... containing ninety acres in two parcels ye greater of which two parcels is bounded on ye North by land in ye possession of Henry Bull.*

That property built about 1640 existed on Spring Street in Newport until a short time ago when it burned down. It can be seen on the Internet at www.dinsdoc.com/weeden-1-3.htm. A picture of that house is included in chapter three of the book *Early Rhode Island* written in 1910 by William B. Weeden. That house of clapboard, stone, and wood is probably similar to the house of Bull's son, Jireh.

In 1657, the Narragansett country was bought up in equal shares of land known as "the great Pettaquamscutt purchase." The tract was obtained for a mere 16 pounds. It covered all that country now included in the townships of North and South Kingston. Chiefs of the Narragansett tribe sold the land, not as a grant, but as a purchase. This purchase of Pettaquamscutt opened up new country, and gave employment to many new settlers, as well as Negro slaves who were beginning to be brought into the Colony, and about whose holding laws were enacted.

The sparse settlers wanted a garrison house. New England garrison houses were mainly private dwellings, but they also served as community shelters in times of danger. Most houses appeared to be one- or two-story farmhouses with modified walls of thick logs, stone, or brick. The latter two materials were valued for their fireproof qualities although used primarily only in the first floor. The second story usually overhung the first as a defensive advantage. The garrison house also featured smaller windows or musketry loopholes or portholes with shutters.

Quakers William and Elizabeth Bundy purchased two hundred acres of land in what is now South Kingston, Rhode Island, May 1, 1663. Twenty acres was laid out as a house lot, "and he hath built upon it." Evidently, William Bundy remained in possession of the land only five or six months when he sold it on October 27, 1663, to Jireh Bull, the "Fearless Quaker" who was one of the first Quakers to settle in Narragansett County, Rhode Island. The deed is signed by William Bundy and "X," the mark of Elizabeth Bundy. The deed states that, "I, William Bundy, late of Narragansett, now bound for Carolina do assign to Jireh Bull …"

Jireh Bull purchased this 20-acre house lot stretching up the hill from the river, which formed the eastern boundary. On the south, it was bounded by a lot the Bull family already had. On the north was "a lot granted Rowse {Rouse} Helme." On the west, it was bounded by land "not layd out which said lot had a house upon it." In 1668, Jireh made an additional purchase from the original Pettaquamscutt purchasers, which included 500 acres and another 20 acres for a house lot.

Pettaquamscutt—a few miles north of Point Judith—was where Bull built a small shipyard and promoted a settlement of some 17 artisans. This tract became a historically important place. It is also a site of two prehistoric stone structures—very unusual for Indian endeavor. This spot is one mile south of the huge rock called Treaty Rock where, it seems, the deed to Providence had been negotiated.

Jireh was also a Conservator of the Peace for Pettaquamscutt along with Samuel Wilson in 1669. His property was perfectly positioned for boat launching and building. (One vessel built in the waterway was 120 feet long.) There, Bull built a large house to trade with the Indians or to serve as a meetinghouse and garrison. It measured 65 by 27 feet. In the days of Roger Williams, the Jireh Bull house was one of the few stone structures in the Atlantic colonies.

A bitter conflict arose between Connecticut and Rhode Island as to jurisdiction in the Narragansett country. In 1670, Rhode Island appointed commissioners to proceed to Connecticut to settle it. The commission came to nothing then but appealed again to King Charles and later was successful.

On May 19, 1671, all the inhabitants of the district were present in Jireh Bull's house built in 1663. The Court was called there, and they publicly read his Majesty's charter and letters, and the Commons' orders. Parliament had decided the matter in favor of Rhode Island. The meeting included Jireh Bull, Samuel Wilson, John Porter, Thomas Mumford, John Tift (Tefft), William Hefferman, Rouse Helme, James Edride, Samuel Albro, Benjamin Gardiner, George Palmer, Stephen Northrup, William Ayers, George Crofts, Enoch Place, and Christopher Helme. Those all gave their allegiance to his Majesty and promised fidelity to this Colony.

As time passed, realizing that the white invaders were overcoming Amerinds both through guile and disease, the Indians attempted to form an alliance of survivors of disease-ridden tribes of New England to drive out the whites. Colonial authorities soon suspected this. Metacom's brother (called Alexander but whose Indian name was Wamsutta) was invited to a banquet of settlers at Plymouth, and became ill. He died and indications abound that his British hosts had poisoned him.

His brother, King Philip (Metacom/Metacomet) arose with many Indian allies to avenge him. Rhode Islanders were entirely out of sympathy with the other colonists who had been fighting Indians. They had long been on friendly terms with the powerful tribe of Narragansetts and lived in their midst. When the Narragansett sachem, Canonchet, proved false to the other whites, the Pettaquamscutt settlers refused to take up arms.

However, records suggest that even though the Narragansetts were reluctant, they allied themselves with the Indian cause in 1675 and joined other tribes raiding throughout southwestern Rhode Island.

Jireh Bull's very house would be the scene of a tragedy that caused The Great Swamp War of 1675 in Pettaquamscutt. The colonials needed a garrison. Captain Waite Winthrop, July 9, 1675, described Bull's blockhouse to Connecticut Governor John Winthrop, as a good place to:

> ... *quarter his troops at Jireh Bull's where there is about 16 of the neighbors, it being a convenient large stone house with good stone wall yard before it which is a kind of small fortification to it.*

The Indians attacked the Bull garrison on December 16, 1675, burning it and killing about ten men and five women and children. It is speculated that the Indians stole up in the night and fired the western end of the outbuildings and with the wind strong from the west, the fire soon reached the house. The Indians tomahawked the inmates as they rushed out, except that two escaped. They may have been Indians and escaped in that way. Others said that two white settlers, the

Eldred brothers, escaped by disguising themselves as Indians. Soldiers arrived the following morning to find the ashes and dead bodies.

When I first read of this escape by the Eldred brothers I was immediately struck with a sense of bondage to time. For the episode compares closely with the narrative of the Vinland sagas and certain other unrelated Viking folklore. It might be recalled that when the "noisome" device had been hurled over Karlsefni's palisade, some or all of the Norse pioneers made what seems a panicked retreat toward some "... cliffs that they knew ..." which were to the north and near the river.

First, in the unrelated saga which I recall here from memory, it seems a Viking family somewhere in Norway incurred the enmity of another nearby family to the extent of a protracted feud. This happened to be a way of life in those times in Scandinavia as it was for long among the Scots. An attacking party had laid siege to a stone residence and a battle ensued with the ultimate end of a fire being set to the structure. One corner of the house had been breached and one hardy male family member leapt through this breach which was engulfed with dense smoke. Outside was a pile of chickweed, which he also hurdled. He then ran along a plank across a stone wall. Even though his enemies were in close proximity, for some reason besides the dense smoke—perhaps as a "chimera"—they did not see his escape and believed him among the dead found among the embers. In great story telling tradition, this saga describes the dramatic escape, the hero's ultimate recovery of fortunes and his eventual return to wreak spectacular vengeance and horror upon those who believed him a ghost.

The Eldred brothers also seem to have made somewhat of a similar escape away from a burning stone house but in their case they were seen and pursued. At a place called Indian Run Brook, one of them turned upon their pursuers and managed to kill one before running on. Before his turnabout he had run just about a mile to a point above Pettaquamscutt Rock. I wonder at the coincidence of Freydis Ericksdottir making the same run in the same direction and near the same distance where she also turned upon pursuers. It does not strain imagination to suppose that these two "last stand" places may have been in the same vicinity.

I suggest that it is a bit more than just a historical coincidence. Analyzing the terrain in an attempt to formulate just how the Viking battle had transpired if it had taken place here, there arose some curious questions and the main was how could I explain why both escapes were to the north. For the Vikings the answer is clear—there was an ideal defensive place there. But why did the Eldred brothers run in that particular direction? It might have been the same reason, which has to do with the terrain. This is something which only detailed topographic maps can indicate, and those not so good as familiarity with the ground itself.

There are constraints as to travel to and from Jireh Bull's and Leif Erickson's home sites. In its early days, it was always approached by water. To the east is the river and also a steep down hill slope. The river at this point must be swum and no one in full flight will yield altitude willingly, for it speeds escapee and pursuer in equal measure. Sooner or later, the loss in altitude must be regained and in this event the advantage is all to the pursuers. So, we can imagine that a human quarry, at least, will run along a ridge rather than ascend or descend.

To the west is a large area of swamp and marshy terrain so difficult that there are few roads or trails through it even today. Topographic maps show the interior as a jumble of small hillocks. This area is the northward extent of the glacial rift paralleling Narragansett Bay from Great Salt Pond. Pettaquamscutt hill rises steeply west and then slopes gradually down to this marsh where an escapee is at a decided disadvantage. No one goes in that direction and neither did I, who felt topographic maps sufficient to the task. We also might ponder here at the coincidence of the attacking native's approach by boats in both incidents.

(I make claim to have not failed altogether, for over the years I have tried to penetrate that area even in those halcyon days when I felt that there was no such thing as terrain so impassable as to thwart my person and my motorcycle from free passage. It sure enough is rough country in there. On the other hand, in olden times this place must have been extremely prolific in grapes, berries and a wide diversity of wildlife—Vinland the Good. Perhaps I should add that there is now a small city, Wakefield, which developed over time at the head of Great Salt Pond and between it and the marsh.)

A southerly direction would also be an unfortunate choice for in that direction lies the cul-de-sac of Point Judith peninsula where the game is to the hunter and not the hunted. No doubt about it, north is the way to go to depart Pettaquamscutt in alarm or languor.

Now to return to the Indian attack, three days later, December 19, 1675, a meeting of all the colonial troops was held in Pettaquamscutt. This was the first regular American army ever collected, under the command of General Josiah Winslow, Governor of Plymouth. The place in which the Narragansetts were to be sought was in what is now the town of South Kingston, a few miles northwest of Bull's house. The Indians were 3,500 and were strongly entrenched on a hill surrounded by swamps. The colonists numbered 1,000 men.

The militia from Massachusetts congregated at Roger Williams' old post in Wickford, and marched overland through Rhode Island to where many Narragansetts had a winter encampment. The English militia, before daybreak on December 19th, came upon the Indians despite ice, snow, and terrible swamp conditions. The militia attacked the palisaded encampment, located on an island in a swamp. After a bitter fight, they overcame the wall, entered the camp, and slew all who could not escape. This was something in the order of 400 to 900 persons, mostly old men, women, and children. They burned over 600 wigwams and the screaming and horrified women, children, elders, and warriors died at the hands of the English. This seems to have been the very first American engagement using a calculated genocidal policy.

Earlier, during King Philip's (Metacomet's) War, Joshua Tefft, son of John Tefft who met at Jireh Bull's blockhouse, had married an Indian woman and had been with the Indians in the winter of 1675. Vikings were not the only ones to interbreed with natives. It was said at his "trial" that he had actually fired upon the British 20 times while with the Narragansetts during the Great Swamp War, and took the scalp of at least one Englishman.

On January 14, 1676, he was captured and brought to Providence. In a letter of that date written by Roger Williams to Governor John Leverett of Massachusetts the tale unfolds:

> *This night I was requested by Capt. [Arthur] Fenner and other officers of our town to take the examination and confession of an Englishman who hath been with the Indians before and since the fight. His name is Joshua Tift [Tefft] and he was taken by Capt. Fenner this day at an Indian house half a mile from where Capt. Fenner's house (now burned) did stand.*
>
> *He was asked how long he had been with the Narragansetts and answered 27 days more or less. In answer to the question how he came among them, he said he was at his farm a mile and a half from Pettaquamscutt where he hired an Indian to keep his cattle, himself proposing to go to Rhode Island but the day he prepared to go a party of Indians came and told him he must die. He begged for his life and promised he would be servant to the Sachem for life, and his life was given him as such slave. He was carried to the Fort where were 800 fighting men. His 8 cattle were killed. He said he was in the fort and waited on his master the Sachem, till he was wounded, of which wound the Sachem died 9 days afterward.*

Joshua's story was not believed and on January 18, 1676, he was executed as a traitor. The 1647 Rhode Island statute read:

> *For high treason (if a man) he being accused by two lawful witnesses or accusers, shall be drawn upon a hurdle unto the place of execution, and there shall be hanged by the neck, cut down alive, his entrails and private members cut from him*

and burned in his view; then shall his head be cut off and his body quartered; his lands and goods all forfeited. [And one wonders at Indian "barbarity."]

Joshua had a 9-year old son who was adopted and brought up well by three guardians, one of whom was Jireh Bull. Tefft became the only Englishman to be executed in this style (drawn and quartered) in America.

For a time, Narragansetts still engaged in small raiding parties at long distances. In spring, after the terrible December Great Swamp Massacre, on March 26, 1676, Canonchet's warriors massacred 65 colonials and 20 Indians led by Captain Michael Pierce. Three days later, on March 29, 1676, they attacked and burned Warwick and Providence settlements but seem to have taken few lives.

Williams, then 77, confronted one party near his burning house and entreated for peace. In a letter (held by the Rhode Island Historical Society) to his brother April 1, he described the incident. I have modernized the spelling and punctuation for easier reading because it clarifies the kind of relations the Narragansetts had with Williams and the whites.

> *I asked them why they assaulted us with burning and killing who ever were kind neighbors to them (and looking back) said I, this house of mine now burning before mine eyes hath lodged kindly some thousands of you these ten years. They answered that we were their enemies joined with Massachusetts and Plymouths [enemy Indian tribes] entertaining, assisting and guiding of them. And I said we had entertained all Indians, being a thoroughfare town, but neither we nor this colony had acted hostilely to them. I told them they were all this while killing and burning themselves who had forgot they were mankind, and ran about the country like wolves tearing and devouring the innocent and peaceable ...*
>
> *They confessed they were in a strange way. Secondly, we had forced them to it. Thirdly, that God was with them and had forsaken us for they had so prospered in killing and burning us far beyond what we did against them ...*
>
> *I told them they knew many times I had quenched fires between the Bay and them and Plymouth and Connecticut and them. And now I did not doubt (God assisting me) to quench this and help to restore quietness to the land again. They heard and understood me quietly. They desired me to come over the river to them and debate matters at large ...*
>
> *I desired one of them to come over, saying they had been burning all the day on this side and were they afraid of an old unarmed man in the same place. They desired me to open my cloak that they might see I had no gun. I did so ...*
>
> *He said you have driven us out of our own country and then pursued us to our great misery and your own, and we are forced to live upon you. I told them there were ways of peace. They asked how. I told them if their sachems would propound something and cause a cessation, I would presently write by two of theirs to Boston ...*

> *I said they were a cowardly people and got nothing of ours but by cheating, our houses, our cattle, and selves by ambushes and swamps and great advantages, and told them they durst not come near our forts ... I again offered my services in a way of peace ... I told them God would stop them or plague them hereafter except they repented of these their robberies and murders. We parted and they were so civil that they called after me and bid me not go near the burned houses for there might be Indians might mischief me, but go by the water side.*

Williams' confrontation at Providence took place while the few small dwellings were actually burning. This occurred about where the great marble State House now stands.

The fledgling "city" of Providence consisted then of a row of dwellings below a steep hill along the east (opposite) side of a river that passes north there. It was quickly rebuilt after the Indians had torched the main area. The present museum for Roger Williams is along the eastern riverbank, which extends just north of the State House and south to what is now called Fox Point, some 3/4 of a mile.

Prior to Great Swamp assault, the size of the Narragansett War Party is significant for it demonstrates the high degree of health enjoyed by the tribe, which is also an inherited trait. At this time, 1676, most New England tribes were reduced from their previous numbers. From what I can gather, during the whole of King Philip's War, no other tribe in New England was able to muster as many as 80 warriors at any given time. The colonial militias, mustered by towns and allied townships, sometimes numbered as many as 150, generally for temporary service only. The key to victory by the colonials lay in logistics. The Indian food supply collapsed in spring of 1676 since they were unable to plant, and that spelled the end of the war.

The strength of the Narragansetts had been irreparably broken in the Great Swamp War and massacre, never to return. The war was extremely destructive to both colonists and Indians and has been called the first and the worst (proportionately) Indian war in America north of Mexico. It lasted a year and ended in August 1676, with the death of King Philip near Bristol, Rhode Island—shot by an Indian mercenary in colonial employ. In fact, it was said that not one local Indian leader alive in 1670 still drew breath in 1677, which gives some hint of how the English built their empire by collapsing native cultures.

At the defeat of the Indian confederation, some Indians escaped, large numbers joined other tribes further west, some were slain, and some were enslaved and sold as far away as the West Indies. A few hundred cooperative colonial survivors were left with a legacy of religious freedom, which may have been not solely a result of a new tolerance, but a fear of consequence from a powerful adversary.

About the Bull property, as time went by Jireh Bull and his family rebuilt a home in 1684 near the one that was burned down. It should be added that Jireh Bull was not killed in the massacre but came back to live on the property. The remains of that last house can be seen in the paving stones overgrown by weeds. There are seven fieldstone marked graves northeast of the site of the Bull Blockhouse on a knoll. The property appears to have been deserted by 1729 according to a town survey. The site has remained of interest to historians of the area. It is of interest to me, as well.

I have always wondered whether Verrazzano's comments about "refugio" referred to the Jireh Bull property in such an appropriate spot to be seen by Viking mariners as they were taken on a tour of the environs. This property was excavated and showed older buildings underground, which have never been further, explored. Let me go into the details of that excavation.

In 1917, the Peckham family farmed the land and allowed Thomas G. Hazard and the renowned architectural historian Norman Isham to excavate the garrison site. It comprises two complete and one partial structural foundations. Artifacts recovered from this excavation are with the Rhode Island Historical Society.

The excavation report was entitled "A Preliminary Report on the Excavations at the House of Jireh Bull on Tower Hill in Rhode Island." It was issued at the General Court of the Society of Colonial Wars in the State of Rhode Island and Providence Plantations, by its Governor, Henry Clinton Dexter, Esq., and the Council of the Society, December 31, 1917, and printed for the Society by E. A. Johnson & Co. Here are some excerpts from that report:

> *Part way up the eastern slope of Tower Hill on that portion of the "Bull-Dyer farm," which is now owned by Mr. Samuel G. Peckham, there has been for many years a series of mounds, betrayed as stone heaps by the outcropping fragments, and marked, in part, as a rectangle by an old growth of buckthorns. The spot thus indicated has always been the traditional site of what is generally called Bull's Garrison or Block House, which was burned by the Indians December 15, 1675, and which, though probably not originally intended as a fortification, did serve as a refuge for 17 of the neighbors, only two of whom escaped the savages ...*
>
> *About ten feet west of this first house, we discovered, by trenching westward, another building, even larger than the first, which we have indicated by B on the plan. The south wall of part of this was in line with the south wall of the eastern house, but the north wall was about four feet north of the northern wall of the Building A.*
>
> *This new building proved to be divided into two rectangles, an eastern and a western, by a heavy partition wall. The western rectangle, again, was nearly divided by a mass of masonry into two others ...*

> *South of the house B, about 20 feet away, was a small structure with heavy walls. This building, which we have called house C, is best described by the plan. It forms three sides of rectangle and measures 16 feet from south to north. It thus has two side walls and a back wall, and against the back wall is a fireplace of which the hearth, with the foundation thereof, has long ago disappeared. South of the fireplace is a place for the stairs or the ladder which served as a stairway ...*
>
> *No well has yet been found. It is almost certain that one existed in the enclosure. A spring still flows several hundred feet to the west, another at the northeast and still another as the southwest; but all are too far away.*
>
> *For years, the buildings must have served as quarry for the farm. It is known that they were still used 80 or 100 years ago for the building of stone walls ... The outer wall of which Winthrop speaks was probably the first to go. Then the stones from the others were taken until the masonry was cut down to the level of the ground where it was soon covered by earth and grass.*
>
> *It may be in order now to make some suggestions as to the history of the various buildings the ruins of which we have unearthed ...* **That there have been three houses on the land is evident. Possibly, there was once a fourth.**
>
> *The first house excavated, "A," is the latest. It was probably built after Bull came back to the site when the war was over, or by his son after Jireh's death in 1684 ... The house which Capt. Waite Winthrop [in charge of English troops in Connecticut] saw and described was what we have called the western building and have designated as "B" on the plan ... It is possibly the house sold to Jireh Bull by William Bundy, October 27, 1663 ... The south building {C} may be the oldest on the place. It looks to be a stone fireplace end for a small wooden house, such as the Carr house was on Conanicut Island before it was rebuilt. More excavation, however, has yet to be done at the east of this fragment ... (Emphasis added, FB)*

Only more excavation might determine whether the other structures or artifacts on the property could possibly be as old as 1000 A.D., since the spot is of interest to any seaman entering the estuary from Narrow River. It is close to shore and the elevation overlooks mooring sites.

That report stated that no well had been found. There is a spring there now. It probably does not always flow. It may have flowed in 1000 or 1660 but not in 1917. The old spring once had a small structure over it, a "spring house" often used as primitive refrigeration for butter, etc. Its rivulet passed right by the blockhouse. The location of springs would always have been a factor and an indicator of residence site.

What relevance does a Carr house on Conanicut Island have? Caleb Carr bought Conanicut land from Jireh Bull. Perhaps this is why excavators of Jireh Bull's house mentioned that the older house appears similar to Carr's house on Conanicut. Conanicut is now called Jamestown and is an island between Pettaquamscutt and Aquidneck/ Newport Island. Jireh or his father, Henry Bull, may have built both houses at an earlier time.

In 1981, members of Brown University led by Stephen Mrozowski conducted a second excavation of Jireh Bull's property but I do not know all of the results of this dig. In 1981, Mrozowski was working on a Master's thesis in Anthropology at Brown University in Providence. His thesis was entitled, "Archaeological Investigations in Queen Anne Square, Newport, Rhode Island. A Study in Urban Archaeology."

He studied the site mainly looking for evidence of the relationship between Henry Bull (this Henry Bull is a descendant of Jireh Bull) and his black slaves. Mrozowski found a 1762 gravestone suggesting to him that the most significant relations of slaves were with masters. His comments are excerpted:

> *This 1762 marker for Mille and Katharine was commissioned by Henry Bull, Esq., a prominent Newport merchant who listed his name in both epitaphs on the stone grounds remained arranged in family plots. Although the two women died a year apart, Bull chose to commemorate them thriftily on a single stone, mentioning his own name on both epitaphs. Husbands and wives lie in separate plots established by their different masters, underscoring that, in the eyes of Euroamericans, an African American's most important connection was to the master, not to any blood relations.*

Because of the limited nature of his study, we have no more information on the age of the underground structures. How old are the older structures on the Bull property? Are any of them older than the English settlement there? Surely, Jireh Bull and Leif Erickson would share the same objectives for residences. Placement of residences is dictated by natural conditions such as channels, lakes, conditions of shorelines, etc.

In November 1983, the Jireh Bull Blockhouse property was listed on the National Register of Historic Places. The Rhode Island Historical Society now owns this property.

"Refugio" and "Norman Villa"

Jireh Bull's property is also of interest because it lies at the site of the possible arrival of seafarers such as Giovanni da Verrazzano and perhaps the Vikings. Although Bull's homes were not in existence at the time Verrazzano sailed through the Pettaquamscutt River, this would have been an ideal site for earlier structures if they existed. Earlier stone foundations there would have been a likely object of curiosity shown to Verrazzano on a tour.

Some thought Verrazzano's "refugio" and "promontorio jovio" ("refuge" and "majestic promontory") referred to Newport Tower. However, Verrazzano wrote this note: "The harbor mouth, which we called 'refugio' because of its beauty." Three sentences later he wrote: "... on the right side of the harbor mouth there is a promontory which we call 'Jovius promontory'" [or majestic headland or high land].

On a 1527 map by Maiollo, the words "Norman villa" are written some five notations above the word "refugio," which places the "villa" closer to New York. On Verrazzano's map of 1529, only "refugio" is noted.

Some people use these references to find beautiful physical features at the mouth of the harbor for "refugio." Some use these words to look for a mansion or estate or prominent buildings such as Newport Tower or my proposal to search the old buildings beneath Jireh Bull's property for the "Norman villa."

The Newport Tower has received much interest since Carl Rafn's publication, which suggested that the tower might be of Norse construction. In 1949–50,

William S. Godfrey was working on his doctoral dissertation for Harvard. As a senior archaeology student, he supervised the dig using other students. His dissertation was entitled: *Digging a Tower and Laying a Ghost: The Archaeology and Controversial History of the Newport Tower* with which he achieved his Ph.D. in 1952.

His findings were that the tower was indeed built by local stonemasons for Benedict Arnold around 1675 as a mill but possibly to double as fortress against Indian attack, which would account for a fireplace on the second floor. All artifacts that were found, including coins and pieces of millstone and fragments of a clay tobacco pipe in the foundation, were English colonial and could not be called anything else. He went down to a level of four feet below the tower to the bottom of the column foundations. He examined the mortar, loam, and fill material as well as artifacts.

His final report was "The Archaeology of the Old Stone Mill in Newport" published in 1951. He began his report with these sentences that captured the gist of the problem:

> *The origin of the old stone tower in Newport has been an important problem in the ancient history of American archaeology for something over a century. The argument swirls around two questions. Who built the tower? When was it built? There are two possible answers to the first of these questions. The second question can readily be answered if the first can be solved. Either persons unknown at some time in the pre-Colonial period built the tower; or it was erected in Colonial times shortly before it is first mentioned in the documents.*

After revealing the colonial origins of the stone mill, many wrote in scientific journals to criticize his methods and his findings. He finally responded with the article "Vikings in America: Theories and Evidence" in 1955. He called himself a "disillusioned Vikingist" and summarized the pros and cons of the subject in spite of a flood of pro-Viking affirmations. He described the main arguments of those affirmations and called them vague and unscientific. He thought that the Norse presence was farther north and that the Skraelings might have been Eskimos.

The results of a recent one-month dig of the stone tower at Touro Park by Chronognostic Research Foundation of Tempe, Arizona, was described in an article in the *Newport Daily News* on November 24, 2006. To write "Touro Dig Comes Up Empty," Reporter Sean Flynn talked with Janet F. Barstad, historian and president of Chronognostic. She told him that despite finding nothing sensational, they only came to "find what was here."

Flynn interviewed Dan Lynch, an archaeologist with Soil Sight LLC, who had used ground-penetrating radar before the dig to locate anomalies. Lynch said that they found no evidence of human presence on the site before Colonial times. Barstad added that they also found no evidence of Native American use of the property, either. Another archaeologist, Joyce Clements, from Gray & Page, Inc. told Flynn, "The ground does not lie to us."

The archaeologists found pottery, glass, buttons, pipes, and pipe stems that could be dated. Barstad added that research had revealed that some 75 to 100 stonemasons were available at the time of construction; that is from 1650 on.

The problem with the 2006 excavation is that restrictions were put on it by the city of Newport. They wanted little disruption to this popular tourist site and gave a time limit of 30 days, extraordinarily quick for such an excavation. Many will question whether that project was too limited to conclude that it was only some 350 years old.

Roger Williams and a Viking Legacy for America

Will and Ariel Durant wrote about the contributions of the Nordics in their 1968 book, *The Lessons of History*.

> *Everywhere the Nordics were adventurers, warriors, disciplinarians ... through intermarriage Nordic Englishmen colonized America and Australia, conquered India, and set their sentinels in every major Asiatic port ...*
>
> *All deductions having been made, democracy has done less harm, and more good, than any other form of government ... In England and the United States, in Denmark, Norway, and Sweden, in Switzerland and Canada, democracy is today sounder than ever before.*

I would like to add another legacy. The renegade colony of Rhode Island had long been a haven for, among others, Quakers who in these early days were quite numerous in the colony. Their relations with the Indians here and in Pennsylvania were governed by their pacifist beliefs, and in the main, seem to have been the key to much less conflict in these two colonies than elsewhere. Most Rhode Islanders survived, whereas a neighboring settlement, Rehoboth in Massachusetts, was virtually annihilated and just about all of their farms burnt out.

Quaker tolerance is based on the belief that Christianity is the true religion and that Christ is divine. God will, in his own good time, guide everyone to

Christianity. This, and pacifism, engenders a philosophy of patience—perhaps more patience than most of us can muster. Quakers are peaceful, it is true, but just a bit short of tolerant.

Roger Williams, while tolerant, had not been a good friend to the Quakers in Rhode Island. As he developed experience and knowledge, he and other independent-minded residents came to accept just about every type of religious thought by the time he entered Indian Territory and government. His brother had resided in Turkey for a time, which may have given even Roger just a taste of Islam.

This was a time when nearby states were banishing people for minor differences in religious beliefs. Mary Dyer, best friend of Anne Hutchinson, had recently come to Rhode Island with her husband William Dyer, Henry Bull's dear friend and co-purchaser of Rhode Island land. (Later, the middle-aged mother of six was executed in 1660 in Massachusetts simply for protesting religious oppression and admitting her Quaker beliefs.)

Nowhere in Europe or its colonies could tolerance such as this exist at that time. "Lese majeste" or offense against a ruler was still a capital crime and frequently punished. To doubt the Christian God or King simply was not tolerated. Yet here are the Narragansetts who had their own "princes" and noble chieftains with their own rights of capital punishment practicing tolerance unheard of in the "civilized" populations.

They too had an established religion as can be seen in their regular and orderly burial practices. Their statement about peace had a profound effect upon Roger Williams and his fellow Rhode Island colonists. Possibly the concept of religious tolerance as we now know it originated in this most noteworthy of people, the Narragansetts.

A bit over a century after these events, that small band of magnificent and classically enlightened men who created our Constitution at Philadelphia considered the success of those two original colonies, and found the result worth transmuting to a national right to be enjoyed by all, "Congress shall pass no law ..."

The consequences and their formulation into political rights, rare to non-existent at any time and any place even in the present day, strike me as an American treasure. Perhaps it is our major source of strength.

Usually wherever man organizes, he seems to ally with the dominant conservative religious faction. Those alliances are nearly impossible to break even should it be desired. That is because King, God, and country become hopelessly intertwined in the minds of the body politic. The body politic constantly struggles between peace of mind versus changing fortunes and circumstances. Only in the

United States can one choose among various courses of action in good conscience and safety. Our separation of church and State serves us well.

Henry Steele Commager, in his book *Living Ideas in America*, described how the Italian scholar, Francesco Ruffini, praised Williams:

> *Henceforth the noble cause of religious liberty may find one who will develop it with greater vigor of reasoning and more copious erudition, but never one, however fervent a believer, who will excel Roger Williams in breadth of conception and in sincerity of advancing that cause.*

Williams wrote an attack on theocracy while he was in England trying to get a charter for his colony. The attack was called "The Bloody Tenant," and had 12 points. I will excerpt two of his important points:

> *Firstly, the blood of so many hundred thousand souls of Protestants and Papists, spilt in the wars of present and former ages, for their respective consciences, is not required nor accepted by Jesus Christ the Prince of Peace ...*
>
> *Eighthly, God requireth not a uniformity of religion to be enacted and enforced ...*

For more information on this remarkable man, Roger Williams, see Mary Lee Settle's book, *I, Roger Williams,* published in 2001 by W. W. Norton. While this book is a novel, extensive research is evident, and it appears more like a biography than a novel. It seems to be a quite superior work, is quite successful in yielding insights into the man himself, and is highly recommended.

It deals mainly with Williams' life in England. Only the last quarter covers his life in America and, unfortunately, very little of his life among the Narragansetts. His early years are well covered and explain why this humbly born Briton was able to move among the higher classes and establish the colony that is now "The State of Rhode Island and Providence Plantations."

Until I read her book, Williams' relationship with Lord Coke had been unknown to me, but Settle's description apparently stems from extensive research in England. I bow to her accuracy and diligence. The novel describes the river crossing where Williams is supposed to have said that he positioned his canoe in a small place protected by a sort of rock and spoke to five Narragansett sentinels.

I have remarked earlier that the west side of the Seekonk River had been entirely within the scope of my youthful curiosity and meanderings. It also happens that in later years, but much before engaging in this study, I made my home afloat on the east shore of these same narrows for nearly seven years. In a sense, it was my own Rubicon in transition from landsman to waterman for a period. This is where the crossing apparently took place. It is where successive bridges were built and ferries existed in times gone by. For this reason, the adventures of Roger Williams have been brought home closely to me. I feel that I have made the crossing with him.

I marvel that this narrow crossing place, which was my home for a time, constitutes one of the major historical sites of all time for Americans. Moreover, perhaps someday it will be noted as the absolute dividing line between separations of Church and State. As the peoples of the Earth increase and as we become closer to one another in all ways, conflicts and cruelties stemming from intolerance of others' spiritual views will hopefully pass from the scene. Until and unless it does, human life and mutual regard will be forfeit in ever-increasing aggregates. Periodic violence will be the prevailing human condition and true and meaningful civilization will be forever out of reach.

Finally, I will conclude my own saga. We believe that no other sailors or people had a long enough presence to cause the genetic impact of the Vikings, who resided and traded in the area longer than most. I believe that at least a few interbred with Narragansett Indians leaving the changes I have noted.

Historians hold that since there were no effects of the intrusion at Vinland, that it has no material significance. However, I am one of many who feel that it did, indeed, have consequences and that the voyagers really have left legacies overlooked. Moreover, one of the most important may lie within the Constitution of the United States, the founding Charter of the State of Rhode Island, and the person and history of Roger Williams.

While Williams was an ordained minister, we see that his view on the ultra-strict Puritan thought of the time was mixed in his emotions. He was the man responsible for the inclusion of freedom of religion in the Rhode Island Charter of 1663, which was a very early—perhaps original—appearance in literature and law. He placed it in print 20 years before and it seems to have been influenced by a custom of the Narragansett Tribe.

I have found that the man who lived in the United States' earliest colonial era and whom we so often refer to as a vital resource to this Vinland study is not as well known as he might be. This is an unfortunate situation as he is a world figure whose contribution to American life is immense. Besides this, he was certainly one of the first of colonists who was in intimate contact with Native Americans. In addition, he was one of the first to write about them as human beings. He was to the English invasions what famed humanitarian Bartolome de Las Casas had been to the Spanish. Therefore, I will end by describing him in some detail.

Roger Williams' surviving writings (his papers were burned with his homestead in 1676) are so voluminous and influential that, in fact, his work has had a continual adverse impact on all later studies of American Indians since that time. For the people of whom he wrote were hardly typical Native Americans. They were, indeed, so unusual that they constitute an episode of European History in combination with American. Williams surprises and startles us with his recorded opinion as early as 1643 that Narragansett Indian origins had been in Iceland.

Roger Williams, English commoner, was a classically educated Londoner who arrived an ordained minister at Salem (near what is now Boston) at age 28 in 1631. This is quite early in American history, Plymouth colony having been founded but a decade before.

Massachusetts Bay Colony at that time extended to all of northern New England including Maine and New Hampshire by Royal Charter. Generally, the aspect of European settlements was a thin and barely tangible presence along the coastline with small colonies and outlying farms at Salem, Plymouth, Thames River of Connecticut (actually an outpost of the Massachusetts Bay Colony), Connecticut River, New Amsterdam at the lower Hudson, and maybe even a New Sweden, in Delaware.

All of these can be viewed as essentially coastal or riverine settlements with communications being by boat. There seems to have been another colony not formally recognized or chartered located in Narragansett Bay on the Island of Aquidneck which was in two parts; that at what is now Newport and another at the northern end at Portsmouth. This is a rather large island, comparable in many ways to Manhattan in New York Harbor. Otherwise, what is now the state of Rhode Island was wholly Indian territory with Narragansetts residing on the west side and Wampanoags on the east.

Actually, Narragansett Bay is hardly large enough to separate populations. Nevertheless, the two groups were only rarely, if ever, at war and intermarriage was common among them. A more reasonable outlook is to view them as a "Narragansett Bay Community" with certain distinct variations around the waterway.

As with Manhattan, Aquidneck or (Rhode) Island has a peculiar geomorphology, having a very narrow strait at its northern or Wampanoag end. For this reason, it most probably was accepted, with perhaps some strain, as lands in common between the two tribes—the nautical, monarchal, and militarily powerful Narragansetts with limited homeland territory and the more typical woodland Wampanoags with considerable land from Narragansett Bay to outer Cape Cod.

Some historians claim that Aquidneck Island was entirely Wampanoag, while others aver that it was Narragansett-dominated. Apparently, the unsettled political nature of the island wrought a situation where European exiles and malcontents from other colonies could escape and reside, not only tolerated by the natives but also encouraged, quite likely as possible allies in Indian dealings with other Europeans.

These independent-minded pioneers or renegades were constrained by circumstances to tolerate other beliefs. The oldest synagogue in North America is in Newport—active from then until now. Oceanside Newport until the 1776 revolution was the major city, seaport, and nominal capital of the colony.

Following the creation of the United Colonies or States of America around 1776, the following phrase was inserted in its new constitution: *"Congress shall make no law respecting an establishment of religion, or the free exercise thereof; or abridging the freedom of speech, or ... etc."* That idea was unique and at odds with the prevalent oppressive governmental mode of state religions. It was followed shortly thereafter by upheavals in France, partly instigated by public resentment of the oppressive bond between royalty and Catholic clerics.

On July 8, 1663, over a century earlier, the English Royal charter to the Colony of Rhode Island and Providence Plantations had this to say; having earlier indicated that it was a political experiment:

> *... established in this nation: Have therefore thought fit and do hereby publish, grant, ordain and declare, that our royal will and pleasure is that no person within said colony, at any time hereafter shall be any wise molested, punished, disquieted, or called into question for any differences in opinion in matters of religion, and do not actually disturb the civil peace of our said colony; but that all and every person and persons may, from time to time, and at all times hereafter, freely and fully have and enjoy his and her own judgments and consciences, in matters of religious concernments, throughout the tract of land hereafter mentioned, they behaving themselves peaceably and quietly and not using this liberty to licentiousness and profaneness, nor to the civil injury ... etc."*

"Howard" (King Charles II) signed this when religious persecution was rampant in Europe. The threat and fact of death for colonists was as near as neighboring Massachusetts. Note that "persons" does not exclude the Narragansetts who at that time were possibly most numerous and certainly powerful entities in New England. It is apparently the first appearance of the concept of religious freedom in English law—perhaps any European law, perhaps any of advanced governance law anywhere.

These episodes predated Pennsylvania, which some historians aver was the origin of enlightened and tolerant governance. William Penn (1644–1718) received his Colonial Charter in 1681, two decades later than Rhode Island's. Penn was a convert to Quakerism and was well aware of the Society of Friends and their tribulations in Massachusetts and liberation in Rhode Island.

However, in 1643, Roger Williams wrote of the Narragansetts:

> *They have a modest religious persuasion not to disturb any man, either themselves, English, Dutch, or any in their conscience and worship, and therefore say (Aquiewopwauwash/Aquiewopwauwock). 'Peace, hold your peace.'*

Could it be that we Americans owe the Narragansetts and Vikings a debt for this? Could it be that a cultural trait of Vikings might have held over 1000 years and reappeared in modern law?

Just so.

Afterword

I would like to invite readers to my web site: www.vinlandsite.com. There you may see maps and artistic renderings of some of the things I have described herein. When inserting these addresses into your Internet provider, do not insert the final period in each address below or the web site will not come up. These periods are simply punctuation marks so that I can end each sentence with a period.

The coast of my proposed Vinland as it is today is www.vinlandsite.com/mapnow.htm. The coast of Vinland as it might have been 1000 years ago is www.vinlandsite.com/mapthen.htm. The coast of Vinland in the 1837 Rafn work is www.vinlandsite.com/Rafn%20map.htm. The coast of Vinland as developed by Wave Cleaver is www.vinlandsite.com/wave%20map.htm.

Tracing the course of Leif Erickson is www.vinlandsite.com/mapleif.htm. Tracing the course of Thorvald Erickson is www.vinlandsite.com/maplthor.htm.

The settlement expedition of Thorfinn Karlseffni and Gudrid is www.vinlandsite.com/mapKarl.htm.

The coastal visit of Verrazzano in 1524 is www.vinlandsite.com/mapVerra.htm. The coast of Narragansett tribal lands in 1635 is www.vinlandsite.com/mapNarrInd.htm. The coast and travels of Roger Williams is www.vinlandsite.com/mapRogWill.htm.

The portrait in black and white of Ninigret is www.vinlandsite.com/Peopleone.htm.

The outlet of Narrow River and a replica of a Gokstad ship are www.vinlandsite.com/about.htm. A picture of arrowheads from the Pettaquamscutt area is www.vinlandsite.com/lilnguaone.htm. A picture of artifacts from the proposed area of Hop is www.vinlandsite.com/langpg2.htm.

To view the area with a bird's eye view or tilting your view to see the area in 3-D, use http://Earth.google.com. On Earth Google, you can search for various sites such as Point Judith, Dutch Harbor and Dutch Island, Narrow River, Pettaquamscutt, Newport, etc. Your computer will have to be new enough to install Earth Google. The 3-D feature located in the Tools section will enable you to tilt the view to examine elevated features as you approach land so that you have the same view as arriving mariners.

Another useful Internet site to view Maiollo's map about the Norman villa and refugio and Verrazzano's map about refugio is www.trochos.freeserve.co.uk/newport1.htm.

Recommended from the U.S. Library of Congress is "New England Coasting Pilot from Sandy Point of New York unto Cape Canso in Nova Scotia and all the Island Breton, etc." by Captain Cyprian Southack, dated 1717. This is most interesting, especially to the nautically inclined, as it shows many differences in outlooks of seamen of olden times and erosion over the centuries. Of particular note is the peculiar handling of Narragansett Bay with the upper reaches near absent.

My previously published work, no longer available, of this project's development is "Voyage of Wave Cleaver" in serial from 1988 to 1997 in the U.S. Library of Congress #98660015, ISBN 1046-5839. This represents a history of our progress from very primitive origins to its present state of argument.

Sources

Adam (of Bremen). *Descriptio Insularum Aquilonis.* A.D.1072–76. Online facsimile edition at www.americanjourneys.org/aj-058.

Adams, Jonathan. "Archaeological History of Rhode Island." University of Vermont course synopsis online. 1999.

Abel, Laurent et al. "Genetic Predispositions to Clinical Tuberculosis." *American Journal of Human Genetics* 67 (2000): 274-277.

Asher, Georg M. *Henry Hudson, the Navigator.* (1860) Brookline, MA: Adamant Media Corp., 2001.

Aziz, Mohamed, et al. "Epidemiology of antituberculosis drug resistance (the Global Project on Antituberculosis Drug Resistance Surveillance): an updated analysis." *The Lancet* 368 (2006): 2142-2154.

Bagrow, Leo. *The History of Cartography.* Chicago: Precedent Publishers, 1985.

Barrett, J. H. "Fish trade in Norse Orkney and Caithness: A zooarchaeological approach." *Antiquity* 71 (1997): 616-638.

Beena, K. R., et al. "Lichen Scrofulosorum: A Series of Eight Cases." *Dermatology* 201(3) (2000): 272-274.

Bellamy, Richard. "Genetic Susceptibility to Tuberculosis in Human Populations." *Thorax* 53 (1998): 588-593.

Bolnick, Deborah A. "Showing Who They Really Are." American Anthropological Association Annual Meeting. November 22, 2003.

Brown, Dale M. "The Fate of Greenland's Vikings." *Archaeology Magazine Online,* published by Archaeological Institute of America. (Feb. 28, 2000).

Brown, Fred N. Internet: www.vinlandsite.com.

Chapman, Paul H. "Norumbega: a Norse colony in Rhode Island?" *The Ancient American* 1(6): (1994): 8-11.

Chung, Juliet. "Hispanic paradox baffles medical community." *Los Angeles Times,* 3 September 2006. Republished in the *Arizona Republic* 4 September, 2006.

Clark, George A., et al. "The evolution of mycobacterial disease in human populations." *Current Anthropology* 28(1) (1987): 45-62.

Closs, Michael. *Native American Mathematics.* Austin: University of Texas Press, 1986.

Commager, Henry Steele. *Living Ideas in America.* New York: Harper & Row Publishers, 1951.

Cox, Deborah, et al. "An Archaeological Assessment of the Pettaquamscutt River Basin." *The Public Archeology Laboratory.* Pawtucket, RI, 1983.

Crichton, Michael. *Eaters of the Dead: The Manuscript of Ibn Fadlan, Relating His Experiences with the Northmen in A.D. 922.* New York: Ballantine Books, 1993.

Crone, G. R. *Maps and their Makers.* London: Hutchinson University Library, 1953.

DeCosta, B. F. "Who Discovered America?" *The Galaxy* (1869, July) p. 8.

Dana, Richard Henry. *Two Years Before the Mast.* (1820) New York: Buccaneer Books, Inc., 1988.

DaSilva, Manuel L. M.D "Cantino Spy Letter," *American Portuguese Online Internet.* (2004).

Delabarre, Edmund. "The Runic Rock on NoMan's Island, Massachusetts." *New England Quarterly* 8(3) (1935): 365-377.

Diamond, Jared. *How Societies Choose to Fail or Succeed.* New York: Penguin, 2005.

Durant, Will and Ariel. *The Lessons of History.* New York: Simon & Schuster, 1968.

Estensen, Miriam. *Discovery: The Quest for the Great South Land.* New York: St. Martin's Press, 1999.

Fagan, Brian, ed. *The Oxford Companion to Archaeology.* Oxford: Oxford University Press, 1996.

—*The Great Journey:* The Peopling of Ancient America. London: Thames and Hudson, 1987.

—*Eye Witness to Discovery.* Oxford: Oxford University Press, 1996.

Fitzhugh, William W. *Cultures in Contact: The impact of European contacts on Native-American cultural institutions, A.D. 1000–1800.* Washington, D.C.: Smithsonian Institution Press, 1985.

Fowles, John V. "Telltale Bones." *Archaeology Magazine* 1983(Sept. 26).

Flynn, Sean. "Touro Dig Comes Up Empty." *Newport Daily News*, 24 November 2006.

Gaggliotti, Oscar E. "Evolutionary Population Genetics: Were the Vikings Immune to HIV?" *Heredity* 96(4) (2006): 280-281.

Gebhard, David. "The shield motif in Plains rock art." *American Antiquity* 31(5) (1966): 721-732.

Gies, Frances. "Al-Idrisi and Roger's Book." *Saudi Aramco World*, (1977) July/Aug., pp. 14-19.

Godfrey, William S. "The Archaeology of the Old Stone Mill in Newport." *American Antiquity* 17(2) (1951): 120-29.

—"Vikings in America: theories and evidence." *American Anthropologist* 57 (1955): 35-43.

Goldowsky, Seebert J. *Yankee Surgeon: The Life and Times of Usher Parsons, 1788–1868.* Sagamore Beach, MA: Watson Publishing Intl., 1988.

Greenwood, Celia et al. "Linkage of Tuberculosis to Chromosome 2q35 Loci, including *NRAMP1*, in a Large Aboriginal Canadian Family." *Am J Hum Genet.* 67(2) 2000 (Aug): 405–416.

Hallgrímsson, Benedikt et al. "Composition of the founding population of Iceland: Biological distance and morphological variation in early historic Atlantic Europe." *American Journal of Physical Anthropology* 124(3) (2003): 257-274.

Hansson, Joskim. "General information about Sweden." Lulea University of Sweden online, 1998.

Helgadottir, Gudrun P., ed. *Hrafn's Saga Sveinbjarnarsonar* Oxford: Clarendon Press, 1987.

Helgason, Agnar et al. "Estimating Scandinavian and Gaelic Ancestry in the Male Settlers of Iceland." *American Journal of Human Genetics* 67 (2000): 697-717.

"Indian remains shed new light on syphilis." *New York Times,* 26 Sept. 1983. (Available free on Internet.)

Ingstad, Anne. *The Norse Discovery of America: Excavations at L'Anse aux Meadows Newfoundland, 1961–1968.* Oslo: Norwegian University Press, 1985.

Ingstad, Helge. *Westward to Vinland.* Toronto: Macmillan of Canada, 1969.

James, Peter, et al. *Ancient Inventions.* New York: Ballentine Books, 1994.

Jones, David S. "Virgin Soils Revisited." *William and Mary Quarterly* 60(4): (2003).

Jones, Gwyn. *The Norse Atlantic Saga.* New York: Oxford University Press, 1986.

Kelley, Marc A. "Ethnohistorical accounts as a method of assessing health, disease, and population decline among Native Americans." in *Human Paleopathology: Current Syntheses and Future Options,* edited by Donald J. Ortner and Arthur C. Aufdeheide. Washington, D.C.: Smithsonian Press, 1991.

Levinton, Jeffrey S. "The Big Bang of Animal Evolution." *Scientific American Magazine* 265(5) (1992) (Nov.): 84-91.

Logan, F. Donald. *The Vikings in History.* New York: Routledge Press, 1991.

Lucotte, Gerard et al. "Distribution of the CCR5 gene 32-bp depletion in Europe." *Journal of Acquired Immune Deficiency Syndromes and Human Retrovirology,* 19(2) (1998): 174-177.

Magnusson, M. et al. *The Vinland Sagas.* New York: Penguin, 1965.

Mallery, Col Garrick. *Picture Writing of the American Indians,* Government Printing Office, Washington, 1893.

Mann, Charles C. *1491: New Revelations of the Americas Before Columbus.* New York: Alfred A. Knopf, 2005.

Markham, C. *The Book of the Knowledge of all the Kingdoms, Lands, and Lordships that are in the World, and the Arms and Devices of Each Land and Lordship, or the Kings and Lords who Possess them; Written by a Spanish Franciscan in the Middle of the XIV Century.* London: Hakluyt Society, Second Series, 29, 1912.

McCrum, Robert, et al. *The Story of English.* New York: Viking Penguin, 1986.

McLeod, Roger. "Norse/Gaelic Words in the Language of Tribes along the Eastern Seaboard of America." *American Antiquity,* 7(1) (1941): 89-90.

McNeill, William. *Plagues and People.* New York: Anchor Books, 1977.

MacWhorter, John H. *The Power of Babel: A Natural History of Language.* New York: W. H. Freeman, 2001.

Meltzer, Monte, *et al.* "Cutaneous Tuberculosis" *eMedicine,* WebMD.

Montgomery, James E. "Ibn Fadlan and the Russiyyah." *Journal of Arabic and Islamic Studies* 3 (2000): 1-25.

Morison, Samuel E. *The European Discovery of America. The Northern Voyages: A.D. 500–1600.* New York: Oxford University Press, 1971.

—*Christopher Columbus, Mariner.* New York: The New American Library, 1955.

Mowat, Farley. *The Farfarers: Before the Norse.* South Royalton, Vermont: Steerforth Press, 1999.

Nichols, Johanna. *Linguistic Diversity in Space and Time.* Chicago: University of Chicago Press, 1999.

Ober, Carole et al. "Inbreeding effects on fertility in humans: evidence for reproductive compensation." *The American Journal of Human Genetics* 64(1) (Jan.) 1999.

Olson, Julius E. et al. *The Voyages of the Northmen.* New York: Charles Scribner's Sons, 1906.

Pap, Leo. *The Portuguese Americans.* Boston: Twayne Publishers, 1981.

Pettersen, Franck. "The Viking Sun Compass or How the Vikings Found their Way Back from New York 1000 Years Ago." *Planetarian* 22(1) (1993) March.

Pfeiffer, John. *The Creative Explosion: An Inquiry into the Origins of Art and Religion.* New York: Harper & Row Publishers, 1982.

Philbrick, Nathaniel. *Mayflower: a Story of Courage, Community, and War.* New York: Viking, 2006.

Poser, C. M. "Viking Voyages: The Origin of Multiple Sclerosis." *Acta Neurologica Scandinavia Supplement* 161 (1995): 11-22.

Prat, Jordi Gomes, et al. "Prehistoric Tuberculosis in America: Adding Comments to a Literature Review." *Memorias do Instituto Oswaldo Cruz* 98(1) 2003: 151-159.

Rafn, Carl C. *Antiquitates Americanæ.* Copenhagen: Royal Society of Northern Antiquaries of Copenhagen, 1837.

Ramenofsky, Ann F. "Death by Disease." *Archaeology* 45(2) (1992): 47-49.

Raviglione, Mario. "The Burden of Drug-Resistant Tuberculosis and Mechanisms for Its Control." *Annals of the New York Academy of Sciences* 953 (2001): 88-97.

Rider, Sidney S. *The Lands of Rhode Island and Massachusetts as they were known to Counounicus and Miantunnomu when Roger Williams came in 1636.* Providence, RI. 1904.

—"Bibliographical memoirs of three Rhode Island authors: Joseph K. Angell, Frances H. (Whipple) McDougall, Catharine R. Williams." Providence R.I., 1880. Available on *HeritageQuest Online.*

Robinson, Paul A. "A Narragansett history from 1000 B.P. to the present," in *Enduring Traditions: The Native Peoples of New England,* edited by Laurie Weinstein *et al.* 1994.

—et al. "Preliminary Biocultural Interpretations from a Seventeenth-century Narragansett Indian Cemetery in Rhode Island." In *Cultures in Contact: The European Impact on Native Culture Institutions in Eastern North America, A.D. 1000–1800,* Washington, D.C.: Smithsonian Press, 1985.

Romanofsky, Ann F. "Death by Disease." *Anthropology Magazine* (1992) Mar/April.

Ruhlen, Merritt. *The Origin of Human Languages.* Palo Alto: Stanford University Press, 1996.

Schlederman, Peter. "Eskimo and Viking Finds in the High Arctic." *National Geographic Magazine* 159(5) 1981): 584.

Settle, Mary Lee. *I, Roger Williams.* New York: W.W. Norton Company, 2002.

Skelton, R.A. et al. *The Vinland Map and the Tartar Relation.* New Haven, CT.: Yale University Press, 1965.

Smelser, Marshall. "Medieval Europeans in America: Latest Findings. *JSTOR,* 1(1) 1967: 7-15.

St. Pierre, Mark. *Madonna Swan—a Lakota Woman's Story.* Norman, Oklahoma: University of Oklahoma Press, 1994.

Sutherland, Patricia. "New archaeology research project explores possible early European contact in Canadian Arctic." *Civilization Ca.* Quebec: The Canadian Museum of Civilization. (2000) April 25.

Trigger, Bruce G. ed. *Handbook of the North American Indians.* Vol. 15. Washington, D.C.: Smithsonian, 1978.

Tucker, William F. *Historical Sketch of the Town of Charleston in Rhode Island from 1836–1876.* Salem, MA: Higginson Book Co., 1998.

Turnbaugh, William A. *The Material Culture of RI 1000, a Mid-17th Century Burial Site in North Kingstown.* Kingstown, R.I.: University of Rhode Island, 1984.

Twain, Mark *Roughing It.* New York: Signet Classics, 1962.

Viereck, Phillip. *The New Land.* New York: John Day Company, 1967.

Williams, Robert C. et al. "Individual Estimates or European Genetic Admixture Associated with Lower Body-Mass Index, Plasma Glucose, and Prevalence of Type 2 Diabetes in Pima Indians." *The American Journal of Human Genetics* 66 (2) (2000) Feb.

Williams, Roger. *A Key into the Language of America.* Bedford, MA: Applewood Books, 1997.

Wingate, Phillipa, et al. *The Viking World.* Usborne: E.D.C. Publishing Co., 1994.

Wroth, Lawrence C. ed., *The Voyages of Giovanni da Verrazzano, 1524–1528.* New Haven, CT: Yale, 1970.

Index

A Key into the Language of America xxv, *140*, *288*
Abel, Laurent 202
Adam (of Bremen) 281
Adams, Jonathan 196
Adbeh, Ibn Khordo 9
Albro, Samuel 259
Algonquian 130, 151, 162
Al-Idrisi xxx, 222, 283
Almagest 164
American Lung Association 212
Amundsen, Roald 225
Andreasen, Claus 23
Antiquitates Americanæ 286
Aquidneck 30, 97, 137, 171, 178, 179, 184, 237, 246, 257, 267, 275
Arnald, Bishop 25
Arneborg, Jette 23
Arnold, Benedict 30, 257
Aufdeheide, Arthur C. 284
Ayers, William 259

Baffin Island 19, 149
Bardson, Ivar 31
Barikmo, Howard O. xiv
Barnstable Harbor 156
Barstad, Janet F. 269
Bartolome de Las Casas 274
Bear Island 59
Bearing dial 14
Bellamy, Richard 202
Berglund, Joel 23

Bishop Erik 30, 114
Black Death 31
Blaskowitz, Charles xxxiv, 121
Block Island 95, 97, 109, 110, 111, 120, 121, 125, 129, 130, 145, 178
Blond Eskimos 148, 149
Boar's Head 33
Book of Knowledge xxx, *222*
Boothbay 130
Boston 32, 84, 94, 116, 231, 245, 246, 263, 275, 286
Bradford, William 193
Brattahlid xxviii, xxix, xxx, 24, 25, 46, 50, 57, 58, 78, 168
Brenton, William 30
Bristol 115, 179, 218, 224, 264
Bruegel, Peter 10, 35
Bull, Henry 256, 257, 267, 272
Bull, Jireh xix, xxxiv, 101, 117, 255, 256, 258, 259, 261, 262, 263, 265, 266, 267, 268
Bundy, Elizabeth 258
Bundy, William 258, 266
Butternut 19
Buzzard's Bay 88, 112

Cabot, John xxxi, 218
Cambridge University 133, 164
Canada xxxi, xxxii, 17, 18, 19, 20, 139, 204, 215, 224, 271, 284
Canonicus 138, 173, 187, 190, 236
Cantino, Alberto 29
Canute the Great 114

Cape Cod xix, xxiv, xxxi, xxxiii, 88, 94, 109, 110, 111, 115, 119, 120, 121, 132, 152, 153, 165, 179, 193, 196, 219, 225, 275
Cape Herjolfness 42
Carr, Caleb 266
Cartier, Jacques xxxii, 219, 220
Cathay 218, 220, 224
Chapman, Paul 182
Chappaquiddick 111, 183
Charles V 130, 131
Charles Waine 163, 164
Charlestown 103, 146, 189, 203, 204, 205, 245
Chesapeake Bay 4, 125
Cheyenne 176, 211
Chilmark 183, 188
China xxxiii, 9, 130, 220, 224
Chronognostic 269
Clarke, Jeremy 30
Clausen, James Earl 117
Clements, Joyce 270
Clifford, Barry 110
Closs, Michael 162
Coal 182, 186
Cocumcussok 137
Coddington, William 30, 136, 179
Codex Flatoiensis xxx, 4
Codex Runicus 31
Coggeshall, John 30
Columbus, Christopher xv, xviii, xxiii, xxx, 218, 223, 285
Commager, Henry Steele 273
Connecticut River 88, 157, 178, 179, 231, 243, 275
Coronation Gulf 149
CorteReal, Miguel 131
Cox, Deborah 200
Crane, Joshua 32
Crichton, Michael 10
Crofts, George 259
Cromwell, Oliver 140, 141
Crook, Wilson W. xiv, 199
Culdees xxvii, 221
Custer, George 211

Dagmalastad ix, 49, 86
Danforth, Dr. John 27
Daniel, Glyn 35
daSilva, Manuel L. 29
Davis, John xxxiii, 220
DeCosta, B. F. 131
Delabarre, Edmund Burke 28
Delaware 275
Denmark xxiii, xxix, xxx, 4, 8, 20, 32, 34, 39, 114, 152, 271
Diamond, Jared 22
Dighton Rock 27, 28, 115
Doyle, Sir Arthur Conan 196
Dungeness 52, 108
Durant, Will and Ariel 271
Dutch Harbor 126, 280
d'Estaing, Count Charles-Hector 121

Edgartown 183
Edride, James 259
Eiricksson, Leifur xxii, 44
Elizabeth Islands 88, 94, 112, 183
Erick the Red xxiii, xxviii, xxix, 3, 4, 23, 24, 34, 42, 55, 77, 78, 86, 113, 114, 224
Erickson, Leif x, xiii, xiv, xxii, xxiv, xxv, xxviii, xxix, 3, 4, 5, 15, 17, 43, 44, 57, 59, 61, 78, 79, 83, 87, 88, 97, 100, 103, 116, 119, 121, 187, 198, 199, 216, 222, 225, 229, 256, 261, 267, 279
Erickson, Thorstein 54, 55, 114, 170
Erickson, Thorvald 6, 33, 94, 95, 119, 246, 247, 279
Erlendson, Hauk 4
Estridson, Svend 20
Eyktarstad 86, 114

Fadlan, Ahmed Ibn xxviii, 9, 10, 170
Fagan, Brian 23
Fagundes, Joao Alvares xxxi, 218
Faroe Islands 8, 17
Farragut, David xxii
Fernandez, Joao xxxi, 218
Finland 12

Finnbogadottir, Vigdis 87
Finnboggi 74, 75
Fitzhugh, William 197
Flatey Book 113
Flato 4
Florida xxxi, xxxii, 203
Fox, George 140
Franklin, Benjamin 169, 173, 233
Freydis 5, 71, 73, 74, 75, 76, 83, 110, 154, 256, 260
Frisland 17
Frobisher, Martin xxxii, 220

Gainsborough 140
Gardar xxvii, xxx, 25, 31, 113
Gardiner, Benjamin 259
Gay Head 94, 172, 183
Gebhard, David 29
Gelisson, Thorkel 4
Gibbs, John 169
Gies, Frances 222
Gilbert, Sir Humfrey xxxii, 220
Glob, Peter 35
Godfrey, William S. 269
Gokstad xxviii, 88, 279
Goldowsky, Seebert J. 203
Gomez, Estevan xxxi, xxxii, 130, 131, 219
Gooseberry Island 155
Grantham 140
Grapes 19, 34, 49, 61, 84, 86, 154, 183, 184, 185, 261
Gray, Edward F. 32
Great Swamp Massacre 189, 190, 263
Great Swamp War xxxiv, 173, 190, 259, 262, 264
Greenland xxi, xxii, xxiii, xxiv, xxviii, xxix, xxx, xxxi, xxxiii, 3, 4, 7, 8, 10, 11, 14, 15, 16, 17, 18, 19, 20, 21, 22, 23, 24, 25, 28, 31, 32, 33, 34, 39, 40, 42, 43, 44, 45, 46, 49, 50, 52, 53, 55, 58, 59, 62, 65, 66, 72, 73, 74, 75, 76, 77, 78, 84, 86, 89, 103, 113, 114, 115, 116, 119, 149, 153, 168, 180, 181, 182, 188, 198, 211, 217, 218, 220, 221, 222, 223, 224, 225, 226, 282
Greenwood, Celia 202

Hahn, Albert G. xv, 154
Hallgrímsson, Benedikt 221
Hansson, Joskim 9, 10
Harold xxix, 3
Hartford 175
Hazard, Thomas G. 265
Hebrides 221
Hefferman, William 259
Heimskringla 4
Helgadottir, Gudrun P. 15
Helgi xv, 74
Helluland xxix, 9, 17, 47, 48, 59, 113
Helme, Christopher 259
Helme, Rouse 259
Herjolf 42, 43
Herjolfson, Bjarne 5, 42, 46, 83, 88, 109, 119, 225
Holmes, Oliver Wendell 203
Homer xvii, 107
Hop ix, xxiv, 19, 53, 65, 66, 67, 68, 72, 74, 83, 84, 85, 86, 87, 88, 107, 108, 109, 110, 115, 116, 117, 118, 146, 155, 156, 256, 279
Hore, Richard xxxii, 220
Howe, Admiral Richard 121
Hub 159, 233
Hudson River 84, 172
Hudson, Henry xxxiii, 131, 220, 281
Huron 165
Hyanno, Mary 187

Iceland xxii, xxiii, xxiv, xxvii, xxviii, xxix, xxx, xxxiii, 3, 4, 5, 6, 7, 8, 10, 11, 12, 14, 15, 16, 17, 18, 19, 20, 21, 23, 31, 32, 33, 34, 35, 39, 42, 59, 72, 73, 79, 86, 87, 89, 113, 114, 132, 140, 148, 149, 151, 153, 158, 168, 180, 181, 186, 188, 198, 206, 211, 215, 217, 218, 221, 222, 223, 224, 225, 226, 275, 284
Iliad xvii, 107
Ingstad, Helge 17, 18, 20, 116, 182, 183
Inventio Fortunata 31
Ireland xxvii, xxviii, xxix, xxx, 7, 8, 11, 16, 17, 34, 65, 154, 221, 222, 223, 224, 242

Irish xxvii, xxix, xxx, 7, 8, 113, 114, 217, 221, 222, 223
Iroquois 139, 165

James, Peter 18
Jones, David S. 212

Kalstegg 4
Karlseffni, Gudrid 86
Karlseffni, Thorfinn xxix, 4, 5, 40, 57, 58, 83, 86, 88, 114, 116, 182, 183, 187, 198, 222, 237, 246, 279
Keelness 52, 60, 61, 65, 83, 107, 108, 110, 118
Keillor, Garrison 33
Kelley, Marc A. 197
Kennebec River 130
Kensington rune stone 30, 31, 32
King Magnus 31
King Philip xxxiv, 162, 166, 169, 171, 172, 173, 190, 259, 262, 264
King Philip's War xxxiv, 162, 171, 172, 173, 190, 264
Knudsson, Paul 31

La Fey, Howard 13
Labrador xxxi, 19, 28, 149, 188, 217, 218, 223
Lake Champlain 172
Leifsbudir ix, xxiv, 5, 6, 19, 23, 25, 40, 41, 51, 52, 53, 56, 57, 60, 62, 65, 75, 76, 77, 83, 84, 85, 87, 88, 95, 107, 108, 109, 110, 113, 115, 118, 121, 183
Leverett, John 262
Logan, F. Donald 84
Long Island 94, 111, 125, 137, 157, 178, 179, 185
Lopez de Gomara, Francisco 131
Lord Coke 135, 273
Lutefisk 185
Lynch, Dan 270
Lynnerup, Neils 25
L'Anse aux Meadows xxiv, xxix, 17, 19, 20, 21, 25, 77, 84, 87, 115, 168, 182, 183, 217, 219, 284

MacWhorter, John H. 159
Magnússon, Árni 4
Maine xxxi, 19, 110, 151, 196, 205, 219, 275
Markham, Sir Clements 223
Markland xxix, 8, 17, 43, 47, 48, 59, 113, 188, 222
Marks, John 184
Marson, Ari 222
Marsten Moor 141
Martha's Vineyard 32, 88, 94, 95, 109, 110, 111, 112, 114, 116, 120, 137, 152, 156, 168, 172, 178, 183, 187
Maryland xxxii, 131
Massachusetts xxxiii, 27, 28, 32, 88, 115, 132, 133, 136, 139, 142, 154, 157, 159, 164, 166, 171, 179, 185, 187, 194, 217, 222, 231, 239, 243, 244, 245, 257, 261, 262, 263, 271, 272, 275, 276, 277, 282, 287
Massasoit 28, 136, 137, 138, 162, 166, 169, 178, 179, 188, 193, 236, 237, 246
Mather, Dr. Cotton 28
Mayhew, Thomas 183, 187
McCrum, Robert 150
McNeill, William 212
Mercator, Gerardus 30
Merrimac 130, 222
Metacom 162, 166, 169, 190, 259
Mic-Mac 151
Minnesota 30
Minuit, Peter 151
Miskaweich, Ibn xxviii, 10
Mohawks 147
Monomoy Island 94, 111
Montauk Point 125
Mount Hope 115
Mount Wachusett 136, 157
Mrozowski, Stephen 267
Mumford, Thomas 259

Nansen, Fridtjof 225
Nantucket 94, 95, 109, 110, 111, 112, 119, 120, 137, 157, 169, 178, 179
Narragansett x, xvii, xxiii, xxiv, xxv, xxvi, xxxi, xxxii, xxxiii, xxxiv, 27, 29, 32, 33, 87, 88, 89,

94, 95, 96, 97, 98, 104, 111, 115, 116, 118, 119, 120, 121, 123, 125, 130, 132, 133, 134, 135, 136, 137, 138, 139, 140, 141, 142, 143, 145, 146, 147, 148, 149, 150, 151, 152, 153, 154, 155, 156, 157, 158, 159, 161, 162, 163, 164, 165, 166, 167, 168, 169, 171, 172, 173, 174, 175, 176, 177, 178, 179, 180, 181, 182, 183, 184, 186, 189, 191, 193, 195, 196, 197, 198, 199, 200, 203, 204, 205, 206, 210, 211, 213, 215, 217, 219, 224, 225, 227, 230, 233, 236, 238, 239, 240, 241, 243, 244, 245, 246, 247, 250, 256, 257, 258, 259, 261, 264, 273, 274, 275, 277, 279, 280, 287

Narrow River ix, x, xix, xxiv, xxv, 87, 95, 97, 99, 100, 105, 116, 118, 120, 147, 155, 156, 201, 255, 266, 279, 280

Natick 136, 152

National Library of Iceland xxii, 87

NEARA 32

Negroes 147, 244

Nestor 9

Netop 134, 135, 139

New Amsterdam 103, 139, 151, 152, 242, 275

New Hampshire xiii, 33, 116, 157, 222, 275

New Jersey xxxii, 130, 131

New York xiii, xxxii, 14, 84, 97, 103, 110, 125, 129, 131, 139, 149, 159, 206, 242, 243, 247, 268, 275, 280, 282, 283, 284, 285, 286, 287, 288

Newfoundland xxiv, xxix, xxxi, xxxii, 8, 10, 15, 17, 20, 25, 28, 77, 87, 108, 113, 116, 119, 149, 168, 182, 217, 218, 219, 220, 284

Newport xxiii, xxxiii, 27, 29, 30, 97, 115, 125, 129, 130, 137, 139, 142, 174, 178, 184, 237, 240, 246, 256, 257, 267, 268, 269, 270, 275, 276, 280, 283

Niantic 103, 136, 146, 152, 205

Nicholas, Friar 31

Nichols, Johanna 150

Nielsen, Richard 31, 32

Ninigret xix, xxv, 102, 103, 105, 106, 146, 147, 148, 151, 173, 204, 205, 231, 242, 246, 279

Ninigret II 146, 173

Nipmuc 136, 152

Noman's Land Island 88, 120

Norroena Society 22

Norse xi, xxiv, xxviii, xxx, 4, 7, 8, 12, 17, 18, 19, 21, 23, 25, 29, 30, 34, 40, 43, 53, 54, 66, 68, 69, 71, 74, 77, 78, 85, 86, 87, 88, 94, 105, 113, 129, 148, 150, 151, 152, 153, 154, 155, 157, 158, 159, 160, 182, 183, 184, 186, 187, 213, 215, 221, 222, 247, 260, 268, 269, 281, 282, 284, 285, 286

Northrup, Stephen 259

Northvegr Foundation 112

Northwest Passage xxxii, xxxiii, 149, 220, 224

Norway xxvii, xxix, xxx, xxxiii, 8, 10, 12, 14, 15, 17, 20, 23, 24, 25, 31, 32, 39, 42, 44, 45, 61, 86, 111, 132, 152, 157, 221, 250, 260, 271

Nova Scotia xxxi, 108, 110, 113, 115, 119, 125, 130, 151, 219, 280

Oak Bluffs 183

Ober, Carole 214

Odyssey xvii, *107*

Olafsson, Skulli xv, 153

Oreaefajokul 31

Orkney 8, 17, 45, 217, 218, 223, 281

Ortner, Donald J. 284

Ovid 164

Palmer, George 259

Pap, Leo 28

Parsons, Usher xxv, 203, 204, 206, 284

Pawtucket 97, 136, 137, 179, 231, 236, 246, 282

Pawtuxet 156, 157, 236, 238

Peckham, Samuel G. 265

Pedersen, Rosemary xiv

Penn, William 276

Penobscot River 130

Pequots 139, 142, 146, 165, 236, 243

Pettaquamscutt River xix, xxiv, 87, 88, 89, 95, 97, 101, 116, 118, 148, 155, 156, 157, 177, 200, 211, 229, 255, 268, 282

Pettersen, Franck 14

Pfeiffer, John 12

Philbrick, Nathaniel 132, 166, 188

Place, Enoch 259

Pleiades 128

Plymouth Harbor 107, 110, 115
Pocahontas 103, 164
Pohl, Frederick J. 40
Point Judith 87, 95, 97, 118, 119, 120, 135, 138, 145, 155, 257, 258, 261, 280
Polaris 14
Pope Alexander 131
Pope Nicolas V xxx, 22
Porter, John 259
Portsmouth 27, 136, 138, 178, 184, 237, 275
Poser, C. M. 215
Potlatch 175, 176
Powell, John Wesley 28
Prince Edward Island 109, 130, 219
Promontorio jovio 129, 268
Promontorium Winlandiae xxxii, 17, 94
Providence x, xxxiii, 25, 97, 134, 136, 137, 138, 139, 140, 146, 151, 156, 171, 179, 190, 196, 210, 230, 233, 235, 236, 237, 238, 239, 240, 246, 247, 258, 262, 263, 264, 265, 267, 273, 276, 287
Prudence 97, 136, 137, 179, 184, 236
Ptolemy 164

Quaker 140, 244, 257, 258, 271, 272
Quanopen 162

Rafn, Carl C. 32, 109
Raven-Floki 14
Refugio xxxi, xxxii, 30, 129, 130, 253, 265, 268, 280
Rhode Island ix, xi, xiv, xv, xvii, xviii, xix, xxi, xxiii, xxiv, xxxi, xxxii, xxxiii, 25, 26, 27, 29, 30, 33, 85, 87, 88, 96, 97, 102, 103, 105, 106, 115, 117, 121, 130, 131, 132, 133, 134, 135, 136, 137, 139, 141, 142, 146, 148, 149, 153, 156, 157, 166, 171, 175, 178, 179, 182, 183, 184, 185, 186, 188, 189, 190, 194, 196, 197, 198, 199, 203, 204, 205, 206, 213, 217, 218, 219, 220, 223, 229, 231, 232, 233, 234, 237, 238, 239, 240, 241, 243, 244, 245, 246, 247, 256, 257, 258, 259, 261, 262, 263, 264, 265, 267, 271, 272, 273, 274, 275, 276, 277, 281, 282, 287, 288
RI 1000 148, 166, 167, 168, 171, 189, 197, 202, 203, 206, 288

Risala 9
Robinson, Paul A. 148
Rowlandson, Mary 136, 162, 173
Ruberton, Patricia 197
Ruhlen, Merritt 150
Rumstick Point 157
Rune 30, 31, 32, 33
Runólfson, Bishop Thorlak 114
Rut, John xxxi, 219

Sakonnet River 97, 115
Salem 116, 133, 134, 136, 178, 222, 231, 275, 288
Salmon 47, 84, 100, 116, 240
Sauna 18
Schliemann, Heinrich xvii
Scolvus, Johannes 223
Seekonk River 134, 235, 239, 273
Settle, Mary Lee xxiii, 135, 273
Seven Sisters 128
Shakespeare, William 164
Sherwin, Reider Thorbjorn 150
Sinclair, Henry xxx, 218
Skelton, R. A. 31
Skolp, Jon 223
Smelser, Marshall 183
Smith, Richard 138
Snorrason, Thorbrand 71
Snorre 86, 87
Sokkason, Einar 25
Somers 168
South Dakota 176, 211, 214
South Kingstown 148, 198
Southack, Cyprian xxxiv, 110, 152, 280
Spar stone 15
Spenser, Edmund 164
Squanto 164, 194
Stefansson, Sigurdur xxxii, 17
Stefansson, Vilijhalmur 148
Stine, Anne 21
Straumfjord 61, 84, 108
Straumney 61, 65, 66, 72, 74, 84, 87, 94, 107, 108, 109, 110, 116, 152, 178, 187

Stuyvesant, Peter 151
Styrmer 4
St. Brendan xxvii, 217, 221
St. Lawrence xxxi, xxxii, 30, 218, 219
St. Pierre, Mark 211
Sun compass 14, 15, 16, 286
Sun shadow board 16
Sutherland, Patricia 19
Swan, Madonna 175, 176, 211, 287
Sweden 9, 12, 20, 31, 32, 34, 206, 271, 275, 284

Thames River 88, 243, 246, 275
The Vinland Map and the Tartar Relation xiii, 287
Thorbjornsdottir, Gudrid 55, 58, 105, 146, 174, 187
Thordson 4
Thorgilsson, Ari xxix, 221, 222
Thorgunna 45
Thorhall the Hunter 62, 64, 110, 183
Thule 224
Tift, John 259
Tift, Joshua 262
Tisbury 183
Tiverton 27, 29, 115
Tollund man 10, 34, 35
Tower 29, 30, 97, 100, 115, 129, 265, 268, 269
Trigger, Bruce G. 173
Turnbaugh, William A. 197
Tyrker 113, 114
Tyson, Peter 14

Ursa Major 14, 164

Valdidida 54
Vasquez de Ayllon, Luis xxxi, 131, 219
Vermont 196, 281, 286
Verrazzano, Giovanni da xvii, xxiv, xxxi, 25, 29, 116, 125, 126, 145, 173, 211, 219, 268, 288
Victoria 149
Viereck, Phillip 194
Vikings xi, xvii, xviii, xxi, xxii, xxiii, xxvi, xxvii, xxviii, xxix, 1, 3, 5, 6, 7, 8, 9, 10, 11, 13, 14, 15, 16, 18, 19, 20, 22, 23, 25, 27, 28, 29, 33, 34, 50, 52, 65, 66, 70, 71, 72, 74, 84, 86, 89, 94, 103, 104, 105, 107, 108, 113, 116, 120, 123, 128, 129, 136, 147, 148, 149, 152, 153, 154, 159, 160, 163, 165, 166, 167, 168, 170, 171, 173, 175, 179, 180, 182, 184, 185, 186, 187, 188, 191, 193, 196, 198, 200, 201, 206, 213, 215, 216, 217, 221, 227, 247, 252, 260, 262, 268, 269, 274, 277, 282, 283, 285, 286
Vinland ix, xiii, xv, xvii, xviii, xxi, xxii, xxiii, xxiv, xxix, xxx, 4, 5, 6, 7, 9, 12, 13, 16, 17, 18, 19, 20, 22, 24, 25, 30, 31, 32, 33, 35, 37, 40, 43, 44, 48, 50, 51, 53, 55, 56, 58, 59, 60, 61, 63, 65, 67, 72, 73, 74, 75, 76, 79, 81, 83, 84, 86, 87, 89, 90, 91, 92, 93, 94, 105, 107, 108, 109, 110, 111, 113, 114, 115, 116, 119, 143, 146, 148, 153, 154, 155, 159, 163, 170, 180, 182, 183, 184, 185, 187, 194, 195, 198, 199, 201, 210, 214, 216, 218, 222, 225, 226, 229, 252, 260, 261, 274, 279, 284, 285, 287
Virginia 86, 87, 164, 168

Wallace, Birgitta 18, 20, 21
Wampanoag 28, 134, 136, 137, 139, 142, 145, 157, 162, 168, 172, 178, 187, 230, 236, 275
Wampum 104, 137, 139, 147, 161, 162, 163, 244
Wampumpeague 139, 161, 163, 203
Wamsutta 259
Warren 29, 157, 166, 179
Warwick 140, 171, 190, 238, 263
Weunquesh 173
Wheeler, Sir Mortimer 35
Wigwam 162, 181
Wilcox, Edward 138
Williams, Roger xvii, xxiii, xxiv, xxxiii, xxxiv, 25, 26, 30, 97, 103, 118, 119, 132, 133, 134, 135, 136, 138, 139, 140, 141, 147, 148, 149, 150, 151, 154, 158, 162, 163, 164, 166, 169, 171, 172, 173, 174, 175, 178, 179, 180, 190, 196, 201, 224, 230, 231, 233, 234, 235, 236, 238, 239, 240, 241, 243, 245, 247, 257, 258, 261, 262, 264, 271, 272, 273, 274, 275, 277, 279, 287
Wilson, Samuel 258, 259
Winceby 140

Wineries 184
Wingate, Philippa 34
Winslow, Edward 166, 193
Winslow, Josiah 261
Winthrop, John 26, 88, 106, 133, 146, 230, 231, 242, 243, 246, 247, 259
Winthrop, Waite 259, 266
Wolter, Scott 31, 32

Woods, Peggy 186
Woods, William 159
Wood, Annie M. 32
Wunderstrands 47, 59, 94, 108

Zeller, Otto 115
Zeno, Nicolo xxx, 218
Zheng He 223

Printed in the United States
218849BV00002B/4/P